# MAN'S SEARCH FOR SUSTAINABILITY

*Comparing the most loved and hated insect communities*

**David Auge**

# Table of Contents

# My Goal

My goal is to inject some nuance and specificity into the discussion since I feel like it can be and sometimes is dominated by partisans who want to argue that the details of sustainability are either completely known or too confusing to grasp. My own feeling is one of profound concern. Sustainability is a normative term – one that communicates a value judgment or prescribed behavior. Thus, some things are valued as sustainable and others are not. I have an interest in seeing the next generation grasp the issue of sustainability in a different manner. I feel passionate about being a part of this effort for 30+ years and a little heartbroken about not knowing how to pass along my experience and thoughts to my children. I do wonder about what may be lost by some not having a community-connected belief in developing the story, but I am desirous that a close look at locusts and bees may spark a discussion that is lacking. Finally, I hope all gain a new perspective that, through the stories of bees and locusts, can provoke a fresh look at the idea of sustainability.

 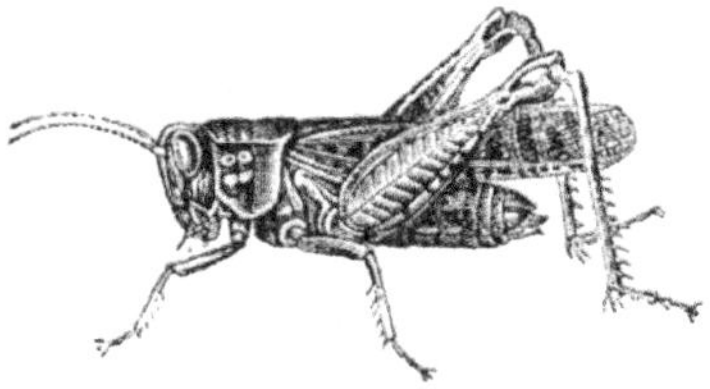

*The human species thinks in metaphors and learns through stories.*

– **Mary Catherine Bateson**

# Introduction: The Story

## Twin Towers

In November of 2001, I attended a luncheon at the West Point Society of Greater Houston. Only two months had passed since the terrorist attacks of 9/11. The men and women talked passionately about the next step—what would our government do? How would we react? When would some of us leave the area for assignments elsewhere? The speaker for the luncheon was an expert on Middle Eastern affairs and had spent years in Afghanistan. The announcement of the topic for this luncheon brought a larger than normal crowd—a crowd interested in the events that would soon unfold before us.

I don't remember all that he covered during the hour-long briefing, but I do remember the theme he used to bring the contrast of this future conflict to our Western minds. "It's simply the understanding of two words," he announced – "it's history." Those words could sum up the reason that the next conflict was going to be so different from what the U.S. had confronted in the past. He noted that the phrase "It's history" to those Americans present meant "it's irrelevant, it's past, it's not a concern," but to those with a Middle Eastern perspective, "it's history" meant almost the exact opposite. Those individuals responsible for the terrorist attacks in New York and Washington, D.C., saw the conflict through an entirely different lens. Instead of 50 or 100 years of conflict, they envisioned the fight from the vantage point of multiple centuries and over a thousand years. If the U.S. was going to enter this war to win, the perspective of "it's history" had to change.

I left the luncheon thinking about the speaker's message and the difference between the two conflicting perspectives of "it's history." Could the U.S. fight a war when we were so quick to forget the history that had brought us to this point?

Around the room were faces creased with the lines that come from participating in numerous historical conflicts. In attendance were men who had landed on the beaches of Normandy during D-day and those who had stood their ground during the attack at the Chosin Reservoir. I looked at the hand of the man who sat near me—gnarled from arthritis and the years he had seen—but still wearing his West Point ring as part of the Long Grey Line. Proudly ready to take whatever part, if just to listen and encourage, that the future may require of him. Men and women who had been part of "it's history" over the past 50 years. Those who had endured and sustained and returned to see another day.

But this is a book about sustainability and insects. Why, at this early point, are we discussing national conflicts in the Middle East? Why bring to the table the willingness of an old man to play his part in history—whatever that may be? Let me share with you the perspective of an environmental engineer with multiple years of working in sustainability at chemical and pharmaceutical production facilities. I find the parallel between war and conflict and the steely resolve of those old soldiers necessarily relevant to the entire discussion. One of the simplest synonyms for sustainability is endurance, and every conflict our nation has participated in during its relatively short 250-year history questions whether we, the United States, can endure when the bullets begin to fly and the decision to fight or run comes to the forefront of every citizen's mind. A common mistake that occurs in the military is to fight the next war with the tactics of the last war. But the perspective of the conflict I was listening to would tie the history of both the last war as well as the future conflict together. "It's history" addresses a worldview difference of living in the now as opposed to history that stretched back into time. But let's go back just 150 years. Remember the beginning of the Gettysburg address?

> Four score and seven years ago, our fathers brought forth on this continent a new nation conceived in liberty and dedicated to the proposition that all men are created equal. Now we are engaged in a great civil war, testing whether that nation or any nation so conceived and so dedicated **can long endure**.

Abraham Lincoln worded it so poetically: "any nation so conceived and so dedicated, whether that nation can long endure." Civil wars tend to destroy from within the fabric of unity necessary to keep disparate groups together. Disparate groups willing to endure. At this critical juncture in our history after 9/11, I was wondering about the resolve of this nation as well. The Twin Towers were burning, the Pentagon had a gashing hole in its side, a fourth aircraft had crashed in Pennsylvania, apparently enroute to the White House. The United States had been "so conceived" as stated in the Preamble of the Constitution by the People of the

United States, in Order to **form a more perfect Union**. A concept of civic improvement and change that would never stop.

A group of leaders (Adams, Washington, Franklin, Jefferson, etc.) were attempting to form something better than what was in the past. A nation based on an idea and birthed through the bloody trials of revolution. This was the "it's history" that gripped me, and as I looked at the men and women in the West Point Society luncheon that day in 2001, it gripped them in a call to action as well. The gnarled hand was still wearing his West Point ring and willing to do whatever his country asked. I believe the story of conflict and men and women willing to act, whatever the cost, creates the narrative one needs to grasp the principal of sustainability and leading a nation during turbulent times. A story that makes "it's history" just a little bit longer than right now—and perhaps forms the virtual spine of a group that would lead a nation through conflict. A group where the responsibility to act was understood as a duty. The responsibility to endure and help this country sustain its effort to form a more perfect Union.

Others have endured over the years in painfully difficult situations. You may find the title of the book, *Man's Search for Sustainability,* familiar. That's because Viktor Frankl's book, *Man's Search for Meaning,* provided me a parallel pursuit. Frankl wrote his book only three months after being released from a Nazi concentration camp in 1945. He suffered much and was still processing how those that had survived had done so. Of those that survived (that sustained to see another day) he noted that they did so because they envisioned:

1. Creating a work or doing a deed;
2. Experiencing something or encountering someone;
3. Or by the attitude and acceptance they took toward unavoidable suffering in this life.[1]

The individuals in the concentration camp developed a personal story which they held so dearly, they were willing to endure almost anything. The meaning or purpose for the soldiers in the luncheon was similar—all could fall under any of the three ways mentioned above. The story of the lives led by those in attendance was prepared to see meaning (and a willingness to help the country endure) by performing a deed, encountering someone new or facing unavoidable suffering. That was the story.

<u>Importance of Stories</u>

In *Future Shock*, Alvin Toffler highlights the importance of stories for understanding where we are, in part, because we have a story that explains how we got here.[2] Story crafts a thread of continuity missing from the stark facts of the moment. Concerning the scenario described above, it merged the faces, young and old, of current, retired, and future military officers into a larger story that was part of the storyline that made our nation's prior wartime efforts successful. Those attending had a story of where we are and how we got here and were ready to act.

But let's consider Toffler's argument again that stories or metanarratives are important. Mark Turner, a researcher at Case Western Reserve University, states it well, "Narrative imagining or a story is the fundamental instrument of thought. Rational capacities depend upon it. It is our chief means of looking into the future, of predicting or planning, and of explaining."[3] Experiencing the world as a series of stories is comforting and orienting and helps create a sense of context. They helps reduce obstacles and impediments by recasting them as bumps along the way to some better place—or at least an end to the journey. Like the soldiers at the luncheon, it creates the narrative for the next step. The step for individuals, whoever they may be, to act.

Do people study metanarratives (an individual's "big picture" of what the world is all about and their part in it) and their impact on decision making? No, not routinely. Studies tend to be quantitative in nature, and they continue to be largely rooted in that which can be measured, whereas metanarrative studies lend themselves more to qualitative inquiry, which is embedded in the interpretive paradigm.[4] The willingness to defend the U.S. or to destroy the World Trade Center stem from a different metanarrative and provides a plot, character, times, and places for the larger discussion of what happens next. By grasping the story, one creates a metanarrative that drives individuals who want to make a difference.

*Details on metanarratives.*

In narrative research, stories are what the inquirer collects, retells, and writes. Metanarratives include some basic elements of narrative inquiry, but, as the word implies, go beyond. They can, for example, take the form of master stories for individuals that form "a comprehensive explanation of all that exists and occurs."[5] The West Point Society members in Houston are an example. In addition to organizing life events more intentionally than narrative, metanarrative requires the integration of historical, sociological, psychological, cultural, and contextual perspectives. From this form of a story the "why" to take action—the motivating force—can be deduced.

Of course, this is not new. Frankl notes that psychologists have various theories concerning how to explain the driving force in human existence. Alfred Adler stated that humanity has the will to exercise power or influence over others, and because of that desire, one decides on the action to take. A person acts, according to Adler, to exert control over the environment and those in that environment. In contrast, Sigmond Freud theorized that all desires and the motivating force of our existence are a basic will to seek pleasure. Human history can be explained as lusting after unconscious memories, thoughts, and urges with the hope to fulfill the id, ego, and superego with pleasure. Sexual impulses and pleasure are the driving mechanisms in Freud's theory.[6]

But a third and, I believe, a more credible approach for explaining humanity's driving force is illuminated by Frankl and *logotherapy*. Mankind's will is to have purpose and find meaning. The drive is making sense out of where we have been and where we are going. It is our own personal story that ties us into the broader history of everything around us. It's a passionate desire to make our history relevant in the larger story of life.

In his 1945 book, *Man's Search for Meaning*, Frankl clearly defends his position. Frankl's book, translated from German to 27 other languages, sold 4 million copies in the U.S. alone. Frankl identified meaning as a central factor enabling people to endure torture and injustice. The experience influenced his belief that the Nazis could take away a lot, but they could never take away one's will—one's ability to have an over-arching story that could make sense of the entire experience this life presents. The will to meaning is the focal structure of Frankl's system of *logotherapy*, according to which a "man's search for meaning is the primary motivation in his life and not a 'secondary rationalization' of instinctual drives."[7] Moreover, Frankl states the search for meaning or purpose in life is closely associated with a personal story. One's answer to "why," which is so important when associating any desire for the present with a relationship with the future and past, occurs here. A person's storytelling, especially to himself, is crucial for his taking action. As humans, pleasure and power are only secondary components of the larger, over-arching drive to make sense of our purpose, to understand the story, and to grasp truthfully the 'why' to sustain anything—ourselves included.

How do we grasp the story or metanarrative of life if such a concept is so important? Douglass Rushkoff, in *Present Shock: When Everything Happens Now*, mentions we can go to the fictional stories of Hollywood. In a simplified fashion, the traditional linear story works by creating a character we can identify with, putting that character in danger, and then allowing him or her to discover a way out. We meet Peter Parker, Luke Skywalker, and Dora the Explorer. Something happens—an initiating event—that sends the character on a quest. Peter Parker wants to avenge the death of his uncle; Luke wants to rescue Princess Leia; Dora wants to get the baby back into its tree.[8] Easy enough.

But the willingness to develop this narrative is under assault. Rushkoff considers the death of narrative as the primary and most critical impact on society's capacity to handle the future. As a result of reality TV we now "generate spectacle without narrative."[9] There are perhaps certain freedoms associated with the collapse of narrative (if one accepts the nihilistic assumption that life is actually meaningless), but the freedom very easily surrenders all of the stories we routinely hear to the basest forms of spectacle and abuse.

Society and individuals all need stories to make sense of our existence and actions and to ground those actions into a large metanarrative. The very motivation of humanity, as Frankl indicates, needs these stories to survive and sustain our culture. Yet, we lack the trust in the stories told and are training our next generation to have lives, experiences, and plans without trustworthy narratives.

*If stories are not trustworthy—what do we do?*

If stories cannot truly come to the rescue of people who no longer have the time or trust required to respond to narratives, what do those people do? What alternatives are there for those who no longer grasp how important stories are? What if stories themselves are incompatible with a presentist culture (an uncritical adherence to present-day attitudes, especially the tendency to interpret past events in terms of modern values and concepts)? Moreover, how do we deal with the trauma of having lost these stories in the first place?[10] As Aristotle states, "When the storytelling in a culture goes bad, the result is decadence."[11] At least on a surface level, the new storyless TV shows appear to support Aristotle's maxim. Why fight, one wonders? Better yet, why do anything?

In fact, without the life sustaining activities foundational for society expressed for us as readily accessible stories, people lose the ability to respond to anything but terror. Rushkoff summarizes this conundrum well:

> If we have no destination toward which we are progressing, then the only thing that motivates our movement is to get away from something threatening. We move from problem to problem, avoiding calamity as best we can. Our worldview becomes increasingly characterized by a sense of panic.[12]

*Panic and a society without a story...*

The loss of story influences Westerners in a unique fashion. The disappearance of a story first incites knee-jerk sensationalism.

During the West Point Society luncheon the difference in the conflicts following 911 would be one of perspective. "It's history." Do those in the Western world have a grasp on a concept of the story as not just relevant but part of true history so as not to pursue knee-jerk sensationalism? How does one think about the world's responsibly to grasp the conflict?

*Aristotle's View*

According to Aristotle, there are two ways of thinking about the world—*mimesis* and *poiesis*. These terms refer to perspective. A *mimetic* view regards the world as having a given order and a given meaning and thus sees human beings as required to discover that meaning and conform themselves to it. *Poiesis*, by way of contrast, sees the world as full of raw material out of which meaning and purpose can be created by the individual.[13] For most of history humanity used the *mimetic* view. There was an over-arching story and we were part of a grand narrative. Unfortunately, the approaches to thinking about the world has changed.

In the West, society has moved away from the *mimetic* view which understood the world as possessing intrinsic meaning to a new view where one creates meaning along the way. *Teleology,* or the explanation of phenomena in terms of the purpose they serve rather than of the cause by which they arise, is thereby attenuated. Why did this happen? Is man a political animal and his changing view of ethics (the critical function of *poiesis*) to be made up as we progress through time or is it that of Christianity, Judaism, or Islam, with its notion that human life in this earthly sphere is to be regulated by the fact that humanity's ultimate destiny is eternal communion with God?[14] The West has moved from seeing life with a purpose we adapt to, in our discovery of truth that matches with the way things are, to creating our own purpose, our own truth, while we walk along the way.

How did this change in worldview occur? Around two centuries ago when more than 50% of the world population resided on farms, farming was utterly dependent on geography and the seasons. Farmers could plough the ground and scatter the seed but exercised no control over the weather and minimal control over the soil. The farmer was entirely at the mercy of the environment. But now, with the advent of more advanced agricultural technology, water, soil science, fertilizers, and pesticides, land can be manipulated to yield more and better crops. Thus, today's world is not the objectively authoritative place that it was 200 years ago and farming is relegated to just a few. Thus, we think of the world more as a box of raw materials which we can almost magically manipulate by our own power for our own purpose.[15] Snap your fingers and you can make it so.

## The Sustainability Discussion

With our changing views society has transfigured the basic *mimetic* narrative that helps make sense of the sustainability discussion. In particular, we have lost some confidence in realistically listening to nature as modern life relies more on data and quantitative reasoning than on the story we are a part of. But why use honeybees and locusts for such a discussion?

First, one of the most desirable of social insects, the honeybee (or *eusocial* insects—those that live in a colony with only some individuals capable of reproducing) provides a society we already admire—and a narrative that could help develop our own. We naturally conceive of virtues in how bees deliberately conduct their colony business.[16] State flags and government emblems promote their positive influence. We learn a lot from this most respected of insects.

Second, the most hated of insects is the one that seems to, as a group, tell humans to get out of the way so it can endure. Stories that begin with locusts never end well. When locusts enter the picture, the farm is no longer sustainable and the community ravaged, so starvation for all is inevitable. Yet, the lessons of the locusts is rich in metahors. "It's history" brings even this insect to life. There are metaphoric lessons to be learned from both insects. As Mary Catherine Bateson stated,"The human species thinks in metaphors and learns through stories." We can learn from honeybees and locust.

## Why metaphors?

A metaphor is when you use a word in place of another to indicate a likeness or similarity. These two communities have similarities to our human society I want to explore. Of course, some scholars warn that we need to avoid "parallelomania" (finding numerous inappropriate parallels) and "parallellophobia" (unnecessarily avoiding the recognition that genuine parallels exist).[17] The stories presented in this book may sway too far either way but since we learn from stories, metaphors will abound. An allegory is a way to take a story and show a hidden meaning. C.S. Lewis stated that "No story can be devised by the wit of man which cannot be interpreted allegorically by the wit of some other man." Therefore, I hope to provide insight from both the most loved and most hated insects in the world through allegory and metaphor. First, we will consider the fascinating world of the bee and locust communities. Second, we will analyze similar functions between these societies and our own. Finally, we will recover knowledge, recover lifestyle, and penetrate the heart and soul of these three societies (honeybees, locusts, and humanity) inhabiting this planet through stories.[18]

### *Fractalnoia*

Rushkoff's book *Present Shock* mentions another failure in any attempt to parallel human communication or experience with disconnected facts. Rushkoff calls it *Fractalnoia*—the now-rampant effort to impose an interpretation of one set of facts on another dissimilar set of facts. *Fractalnoia* is the fourth of the *Present Shock* concerns. Dozens of websites and YouTube videos assert connections and conspiracies between the use of weather balloons and the military, economy and natural disasters, as well as jet emissions and hurricanes. These "make up just a tiny fraction of the so-called conspiracy theories gaining traction online and in other media, connecting a myriad of loose ends, from 9/11 and Barack Obama's birthplace to the Bilderberg Group and immunization."[19] I hope to avoid this. In a fractal, it's not how much you see but how well you see it. I hope to take what can be learned and seen, then apply it. Let's start with an example. A well-known story about Professor Agassiz, the famous naturalist who taught at Harvard in the 19th century, brings the idea of finding something more extensive to light.

## Look and look and look.

> It was more than fifteen years ago [from 1874] that I entered the laboratory of Professor Agassiz … He asked me a few questions about my object in coming, …
>
> 'When do you wish to begin?' he asked.
>
> 'Now,' I replied.
>
> 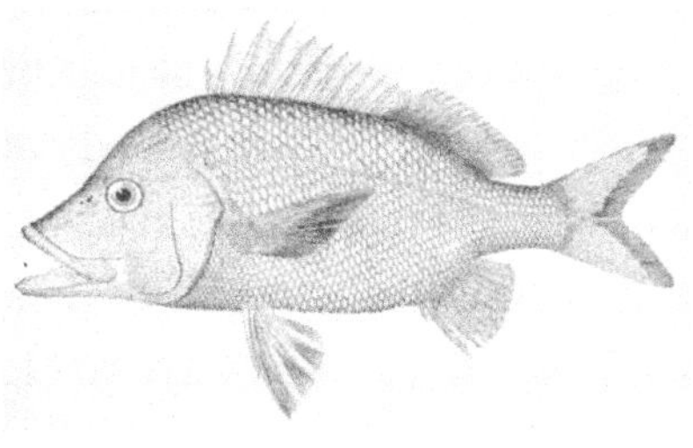
>
> This seemed to please him, and with an energetic 'Very well!' he took from a shelf a huge jar of specimens in yellow alcohol. 'Take this fish,' said he, 'and look at it; … by and by, I will ask what you have seen.'
>
> With that, he left me but, in a moment, returned with explicit instructions as to the care of the object entrusted to me. In ten minutes, I had seen all that could be seen in that fish and started in search of the Professor—who had, however, left the Museum; .... Half an hour passed —an hour—another hour; the fish began to look loathsome. I turned it over and around; looked it in the face—ghastly,

from behind, beneath, above, sideways, at a three-quarters' view—just as ghastly. I was in despair; at an early hour, I concluded that lunch was necessary, so, with infinite relief, the fish was carefully replaced in the jar, and for an hour, I was free.

On my return, I learned that Professor Agassiz had been at the Museum but had gone and would not return for several hours. …Slowly, I drew forth that hideous fish and, with a feeling of desperation, again looked at it. I pushed my finger down its throat to feel how sharp the teeth were. I began to count the scales in the different rows until I was convinced that that was nonsense. At last, a happy thought struck me —I would draw the fish; with surprise, I began to discover new features in the creature. Just then, the Professor returned.

'That is right,' said he; 'a pencil is one of the best of eyes. Well, what is it like?'

He listened attentively to my brief rehearsal …When I had finished, he waited as if expecting more, and then, with an air of disappointment:

'You have not looked very carefully', and he left me to my misery.

I ventured to ask what I should do next.

'Oh, look at your fish!' he said, so for three long days, he placed that fish before my eyes, forbidding me to look at anything else or use any artificial aid. 'Look, look, look,' was his repeated injunction.

This was the best entomological lesson I ever had—a lesson whose influence has extended to the details of every subsequent study; a legacy the Professor has left to

> me, as he has left it to many others, of inestimable value, which we could not buy, with which we cannot part.
>
> Agassiz's training in the method of observing facts and their orderly arrangement was ever accompanied by the urgent exhortation not to be content with them.
>
> 'Facts are stupid things,' he would say, 'until brought into connection with some general law.'
>
> At the end of eight months, it was almost with reluctance that I left these friends and turned to insects; but what I had gained from this outside experience has been of greater value than years of later investigation in my favorite groups.[20]

Perhaps we don't learn sustainability lessons from nature because we don't look, look, look diligently enough to see what others saw. We fail to take the *mimetic* lessons from the world around us and see a general law.

## Defining sustainability.

Perhaps you agree with Agassiz's idea that "facts are stupid things until brought into connection with some general law." I do as well. I see a general law in the narrative or the metanarrative that makes sense of the whole as very important. Sustainability, like the fish being observed, is a bunch of facts in search of a general law. Sustainability measurements are routine tools and methods that attempt to quantify the degree of resiliency of processes, products, services, businesses and so forth. Sustainability is difficult to quantify, perhaps even immeasurable.[21] In fact, that inability to measure sustainability is why observing things closely and seeing the growing narrative already provided in the world is so important. It takes facts and gives them the quality from which we can grasp the entire picture and a general law. For that reason, bees and locusts will hopefully provide you with the opportunity to find a general law.

## The Why Question

I mentioned Frankl earlier as the author of a significantly more influential book—*Man's Search for Meaning*—than what you currently have in your hands. He wrote his book in approximately nine days, only months after his liberation from a concentration camp in Nazi Germany. Although the book details the unfathomable horrors and punishment he endured at Auschwitz, Dachau, and other camps, his primary purpose was the exploration of the will to survive. He addresses an unfathomable oppressor with those that survived and how they found meaning in this life. Nietzsche's quote, "He who has a why to live for can bear almost any how" appears throughout his book.

Frankl presents clearly that man's primary motivation is the desire to discover the meaning of our individual and corporate existence. We want to know why. His experience in the concentration camps exposed that it was not the exercise of power or sexual exploits that drove men to continue. It was the will to have meaning by i) creating a work or doing a deed, by ii) experiencing something or encountering someone, or iii) by the attitude taken toward unavoidable suffering at that moment. Instinctual drives shrivel and disappear under the suffering and pressure of ultimate purposes when your life is slowly being ground into nothingness and whether to sustain yourself for just one more day seems frustratingly impossible.

### Existential frustration

Frankl notes that mankind is unique in its existential frustration.

> Paradise is closed for the man—man has to make choices. He can't depend on instincts (like animals). No instinct tells him what he has to do, and no tradition tells him what he ought to do; sometimes, he does not even know what he wishes to do. Instead, he either wishes to do what other people do (conformism) or he does what other people wish him to do (totalitarianism).[22]

Bees and locusts do not contend with this frustration. It is a uniquely human trait—why do we exist, and what are we to do since we do exist?

<u>Without meaning or a story we exist in an existential vacuum</u>

In our current society, an existential vacuum exists. The boredom we contend with because of an excessive amount of free time is filled with confusion and a failure to pursue the true meaning and purpose of our lives. Here's where sustainability intersects with Frankl's ideas. The pursuit of sustainability is as much a 'why' question as it is a 'how' question. The existential vacuum our society struggles with concerning sustainability contains so many "**how** do we sustain something?" that the critical "**why** should we sustain anything?" seems lost. As a result, conformism (do what everyone else is doing) or totalitarianism (do what other people command you to do) shortcut one's pursuit of meaning and sustainability.

In a way, an evaluation of honeybees and locusts is an evaluation of those societies that have become conformist or submitted to totalitarianism.

*Meaning and sustainability*

The meaning of life (just like sustainability) is not a 'pot at the end of a rainbow' but a continual pursuit. Sustainable is a normative term like beautiful or enduring. A simple definition for sustainability (like a definition for meaning) may never be possible, but the connection with resilience, adaptive capacity, and vulnerability are important. Concerning man's search for meaning, Frankl mentions that the United States should have both a Statue of Liberty on the east coast and a Statue of Responsibility on the west coast. The quest for meaning in life as well as sustainability entail both—the liberty to pursue individual goals and the responsibility to act within the larger goals of the world in which we live.

*How to address the "why" question in life*

Frankl challenged his patients in their pursuit for meaning to—

> Live life as if you were living already for the second time and as if you had acted the first time as wrongly as you are about to act now![23]

The question asked is to urge every individual to live circumspectly. At this point, it is to look at nature. My attempt is not to be a painter but to be an ophthalmologist—to enable one to see the world as it really is. The more one forgets himself by giving himself to a cause or another person to love, the more human he is. Similarly, the more human he/she is, the more likely the 'why' questions asked and the decisions of what to do now will be answered correctly.

## Why study Insects?

As we study animal behavior, we may gain insights related to the study of human behavior and social organizations. George Bernard Shaw, in his play "Man and Superman," states, "Go to the bee, thou poet: consider her way and be wise.[24] In other words, learn from her. Pay attention to these insect communities and the world around you. There is an opportunity when we observe them closely to gain insights we lack. Insight that can be gleaned in some cases from the insects in your own backyard like it has for me. The investigation may even prompt you to build and sponsor a beehive in your own backyard. In fact, beekeeping is exploding as a hobby.

<u>Why is beekeeping participation increasing?</u>

First, it's legal in most jurisdictions. I have beehives in my backyard which is in a residential neighborhood. Very few, if any, ordinances or homeowner associations limit beehives and beekeeping. Second, people are seriously disturbed about environmental and global warming issues and want to do something. They're worried and want to promote a healthier world. Similarly, to those raised in the 60's, beekeepers want to return to the land and learn something fresh. And finally, it's an edgier hobby than knitting scarves or growing zucchini. In fact, it's made up of individuals who are willing to interact with a small insect whose stings can make life miserable, kind of like the "airborne ranger" farmer of the healthy set. These individuals are exceptionally curious, eager to learn and eager to share information—information that provides a look at a society of 20-50,000 inhabitants dealing with environmental stressors every day.

Tying human qualities to bees and locusts

The stories I present have one thing in common; they are projections, connecting humanlike qualities, values, and motivations to bees or locust behaviors that have no real humanness associated with them. Even though our lives are far removed from insect families, their activities will provide an excellent school room. What we learn from bees and locusts provides a lens through which to view ourselves. The key to this effort is not to go too far (I hope to keep the parallels within reason) and an ability to trust metaphors to illuminate human traits (but to go far enough so as not to insult your intelligence).

So I will address sustainability primarily as a philosophical question—how are these insect communities addressing stressors to their existence? How do they allocate work among their community? When a decision that will impact thousands or perhaps millions of individuals needs to be made, how do they make those decisions? With so many messages and so much to communicate, how do these insect communities keep everyone informed? Finally, how do these groups maintain balance with so many competing interests?

So, let's begin and look for those general laws of sustainability from the honeybee and locust.

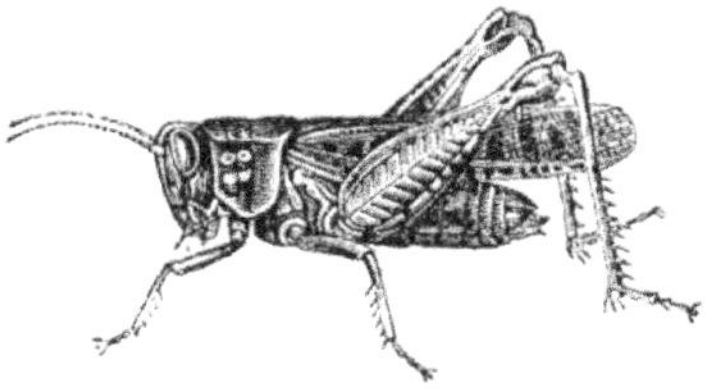

*If the bee disappeared off the face of the earth,*

*man would only have four years left to live.*

– **Maurice Maeterlinck**

# Chapter 1: The Honeybees

A few years ago, I drove down the dusty road of an active farm field near Dunn, North Carolina in search of an apiary selling beehives to local beekeepers. It was early spring and numerous flowers—white, pink, and red—were in bloom. The apiary we visited had well over 160 hives scattered around the perimeter of the farmer's open and recently planted field. Some hives were in logs, but most were in nucs (short for nucleus) or five frame bee boxes. The beekeeper had split his active hives into smaller, saleable units from which a new hive could form.

My son and I had never managed a hive and were concerned about being in proximity to approximately 5 million honeybees. The bees, as they flew past, didn't seem to take notice of us. They flew quickly to their assigned boxes (painted differently so the bees wouldn't get confused) and went inside. My son and I were dutifully outfitted with the latest in beekeeper apparel (direct from Amazon): complete bee suits; gloves; smoker; and hive tool. The veteran beekeeper smiled at us as we donned our equipment to move the purchased hives from his farm to our van. He was a rugged man with a handshake like a vise. He wore no bee suit nor gloves. His 2-year-old grandson, with not much clothing beyond shorts and a t-shirt, followed him everywhere. He too inspected our outfits with a puzzled look.

As we walked around the farm, the farmer proudly told stories of the various hives in tree trunks collected from neighboring areas he rescued. He was stung once on the hand and once below the eye during this tour. He shook it off with a smile and said, “don’t worry, you’ll get used to it.” We laughed nervously.

He confidently described the personality of the bee boxes we were considering purchasing. Some were more aggressive. Others were docile. Some bee boxes produced more honey and appeared to have younger queens. His grandson waived away a few bees and with no shoes on, walked in and around numerous busy hives with no apparent concern. Would we eventually be this cavalier? We wondered…

After paying for our new collection of bees, we gingerly picked up the sealed boxes and placed them in the family van. My son and I kept our suits on as we traveled down the highway back home. Every now and then, a bee, having escaped the hive and now inside the van, whizzed by our faces. Thank goodness for the suits. We looked at each other and smiled. We respected this insect and weren’t going to get into a wreck from a sting while driving. Our adventure as beekeepers had begun.

---

## Apis mellifera

Bees are captivating creatures. More than twenty thousand known species of bees currently live on every continent except Antarctica. Their flower-visiting habits range from omnivorous, taking nectar and pollen

from a wide range of flowers, to obligated relationships, in which a plant species can be pollinated by only one bee species, which in turn visits only that plant's flowers.[25] In fact, 46 universities in 26 states have programs specifically focusing on apiary science.

Currently, ten to twelve honeybee species exist, mostly in Asia, but the best known is the Western honeybee, *Apis mellifera,* (Latin for honey-bearing or honey-gathering bee) which once ranged from Africa to Europe but is now found globally due to human transport of colonies.[26] In the U.S., *Apis mellifera* is the most common species.

Honeybees are not routinely used as competitive team mascots, logos, or names. If even considered, the bumblebee or the fighting bee is chosen, never the honeybee. Perhaps this is the case because of their appearance. Honeybees are decidedly fuzzy, almost teddy-bearish; their bodies are thickly covered with plumose hairs that efficiently catch pollen grains when a bee scrabbles through a flower. They are not the most graceful of flying insects and present one with the idea of a flying minivan (capable of carrying tremendous weight, but not very gracefully) versus the wasp (the corvette of the insect family).

In contrast with the honeybee, nearly all wasps, including the familiar paper wasps and yellow jackets, are predators that kill other insects or spiders (often by stinging) to provide the egg-laying females and their growing young with proteinaceous food. Honeybees, however, depend instead on collective protein-rich pollen from flowers.[27]

What do bees do?

Bees provide honey and shelter for innumerable feral animals. Approximately 20% of all angiosperms (advanced plants) can be pollinated only by bees, while another 45% can be pollinated by bees but also by wind, or other animals, mostly beetles, moths, butterflies, birds, and bats. That is, 65% or 229,000 of the 352,000 species currently known require or benefit from bee pollination.[28]

The bees don't provide this benefit altruistically. There is a motivation keeping the bees on track, and that is their diet. Specifically, the blossoming plants have flowers which provide the bee with both sugar in nectar as well as protein in pollen.

Honeybees do something unique among bees. Floral nectar is carried back to the nest in the worker bees' specialized second stomach, where enzymes are added that break down the complex forms of sugar in nectar into simpler, inverted forms that are easier for bees, and for us to digest.[29] These simple forms of sugar are also better at resisting bacterial attack than more intricate sugars. Other enzymes catalyze the production of tiny amounts of at least two, and probably more, distinct compounds, hydrogen peroxide and methylglyoxal (MGO), which together prevent bacterial and fungal degradation. The foraging honeybees regurgitate the now-processed nectar when they return to the hive, evaporating some of the water by layering it thinly on their tongues and then placing it into cells. Younger house bees stand over the nectar and fan it with their wings, reducing the water content of the nectar to 18.6% or less to protect the honey from yeasts. At this point, the nectar is "ripened" and can be called honey.[30]

Now I am a novice beekeeper with four hives in my backyard and five hives kept on the property where I work. I have, however, preferred the taste of honey in contrast with other sweeteners my entire life. Local honey has an allure of the countryside from which the bees forage that can spur the imagination of all humanity. Listen to this author poetically extol the flavor of a honey crop in Minnesota:

> Here is the taste of two weeks in a particular grove of soaring basswood trees digging their roots into the sheltered floodplains of the Minnesota River Valley. Here is the flavor of a month in the life of a stand of gnarled, wind-twisted horse chestnut trees living by the Blue Earth River. Here is the essence of a scrubby and scrappy stand of red sumac soaking up the sun out by Parley Lake near the town of St. Boni.[31]

Honey flavor can capture the character and nuance of such a valley just like wine.

Did you know…

Among the treasures discovered during the famous 1922 archaeological excavation of King Tutankhamun's tomb was a jar of honey. Archaeologists tasted it (brave souls those archeologists), and to their amazement, found it to be as sweet as, well, honey.[32]

As a result of the honeybee process of reducing the water content of nectar from 60% to less than 18.6% and adding enzymes that break down the complex carbohydrates to simple sugars, honey has been able to last 3000 years and still be honey.

Speculation as to how this novel society works was even present in ancient Greece. Aristotle believed the female queen to be a male king and was convinced that the worker bees developed from some progenitor substance. Aristotle surmised that adult workers were birthed from flowers rather than from eggs laid by the monarch. He did note four life stages of honeybees—eggs, larvae, pupae, and adults—and surmised that worker bees fed the larvae.[33]

FRANCIS HUBER.

Francois Huber, a blind Swiss entomologist from the latter part of the eighteenth century, discovered that the monarch in the hive was a female. Honeybees reproduce by swarming when the old queen and most of the workers leave the nest to seek a new site. The developing queen larvae are left behind and soon emerge as adults to fight other unmated queens to the death until only one is left. The last virgin standing leaves the hive to mate, copulating with ten to twenty male drone bees that catch the queen in flight and then die following their aerial tryst.[34] She never leaves the hive again. Huber's blindness resulted in developing a powerful skill to ask discerning questions of his wife and assistant about what they saw, leading to incredible results concerning the operation of the beehive.

As Huber and others studied the honeybee hive, they began revealing the complexity of this superorganism, the hive, which is not much heavier than a human brain (3.5 lbs). The first hives used

for studies were nothing more than hollow logs hung from trees to catch swarms, with open ends through which early beekeepers could reach in and remove honey-laden combs while also observing the bees at work. While this was adequate for study, it lacked the ordered arrangement we use today. A Philadelphia clergyman Lorenzo Langstroth invented a box with removable frames that resulted in a significant increase in colony productivity and augmented our capacity to observe the complex social mechanism that holds honeybee colonies together.[35] The Langstroth hive is still the standard for the industry to this day.

Now the creation and storage of honey is the primary driver for the organization and effort of the colony, and the amount collected is a key performance indicator of honeybee health. For example, A typical colony in temperate climates requires 130-180 pounds of honey annually to survive. To create a pound of honey, tens of thousands of bees fly about fifty-five thousand miles in total and visit more than two million flowers.[36] The nectar, swallowed and processed in their stomachs, will fill each available cell, be dried by other bees, and capped with wax for access during the winter months.

To survive winter, a crucial trial for any society, a critical supply of the honey collected needs to be made available. In fact, 20 kilograms (44 pounds) are needed for an average hive to last the three to four months of winter. Regarding work effort, given that freshly collected nectar is (on average) a 40% sugar solution and fully ripened honey is roughly an 80% sugar solution, and given that a foraging bee typically brings home a nectar load weighing about 40 milligrams (0.001 ounces), we can calculate that the collection of enough nectar to produce 20 kilograms (44 pounds) of honey requires more than 1 million foraging trips by a colony's workers. And when one also considers the miles flown and countless blossoms visited on each foraging trip, one realizes what prodigious efforts the bees make over summer to sustain their colony through winter.[37]

Some would consider the importance of this valuable commodity, honey, to be the rationale for honeybee management around the world. However, the cost of the production of honey is not the current driver keeping honeybees in the United States. According to 2013 statistics, the cost for a Chinese beekeeper to produce a pound of honey is around U.S. $0.25, while the same pound costs an American beekeeper about $1.25.[38] As a result, the honeybee industry turns not so much on their delectable production of honey but on pollination. In fact, most of the honeybee colonies in the U.S. find their way via truck to California every year. Why? For vacation? No! For the pollination of almonds.

80% of the world's almond production takes place in California. Almonds are the second largest value crop after dairy and the state's number one export, earning approximately $3.8 billion per year. The trees bloom in February; their flowers produce pollen that attracts bees but not much nectar. Almonds are both a high-value crop and one that requires bee pollination to set seed. Growers are willing to pay a premium price to bring 60% of the honeybee colonies in the U.S. to California when the flower of the almond blooms.[39] What a worker!

The ordered structure of a colony of bees

In 1948, Pope Pius XII asked an important question.

> What lessons do the bees give to men? Bees are models of social life and activity, in which each class has its duty to perform and performs it without envy, without rivalry, in the order and position assigned to each, with care and love....Ah, if men could and would listen to the lesson of the bees: if each one knew how to do his daily duty with order and love.[40]

Let's look at the structured order of a beehive.

*The Queen*

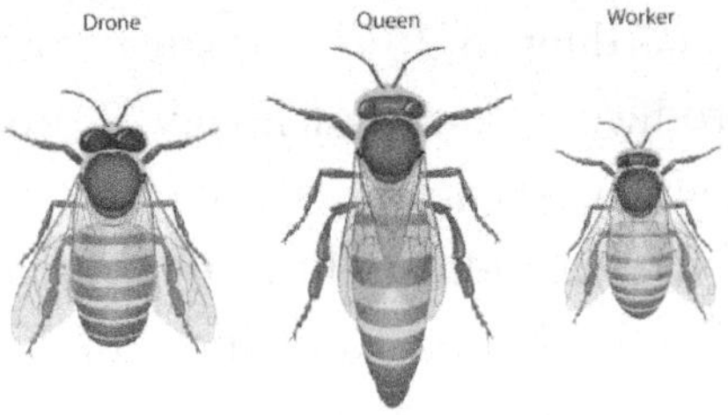


Is the queen a monarch like in human society? No. She influences but doesn't command colony activities. In fact, researchers and keenly observant beekeepers note that life in the hive isn't quite as royal as for human monarchies since governance in honeybee colonies is highly decentralized.[41] In fact, the only decision made by the queen is whether to have a worker, a queen, or a drone. The egg she lays will develop as a female worker if fertilized with sperm held in a sac near the queen's ovaries. Next, the egg will be a queen if fed extra portions of a particularly nutritious diet called royal jelly. If the queen's egg is not fertilized, then a drone results.[42] In the larger scheme of things, this may seem unimportant. However, the number of eggs laid daily (1500-1800) is the very lifeblood of the hive.

If the queen is absent or not quickly replaced because of death, chaos can upend the peaceful order of a well-functioning hive. The workers sense her loss within minutes due to diminished queen odors and begin rearing a new queen within hours. Occasionally the hive fails to raise a queen in adequate time; other workers begin developing their ovaries and attempt to fill her place. Their effort results in excessive numbers of drones resulting in conflict, and anger within the hive (a problem I will discuss in chapter 7). Without a leader every bee begins doing that which is right in their own eye. Internal fighting results in the death of even the most faithful of the remaining bees. A beekeeper's phrase for that situation (worker's laying their own eggs) says it all; the colony becomes hopelessly queenless.[43] One queen per hive is very important.

*Workers*

Of course, even though there is only one queen per hive there are many workers. The hive consists of ~99% female workers, <1% male drones, and a queen. We will spend much of the book discussing the workers and their qualities, so my summary of the key workforce here will be limited.

Worker bees are smaller than both the queen and the drone. In comparison with the other two types of bees, workers have expanded legs with stiff hairs to collect pollen and a concave basket on the hind legs in which to carry the pollen back to the nest. Worker bees ingest nectar during transport and have glands in the abdomen that secrete wax to build a comb. The workers also have a well-developed barbed stinger for defense.[44]

*Drones*

The male component of the hive, drones, are larger than worker bees but not as large as the queen. Drones have no stinger. The largest bee (or close to it) coming out of the hive cannot defend the hive. Approximately 100 drones exist per hive at a time. Next, because of the size of the drone, more food is consumed by them then workers. Finally, their sole function is to mate with virgin queens from nearby colonies. Drones are the brawniest bees, endowed with huge eyes for spying out young queens venturing out on their nuptial flights and massive flight muscles for charging after the queens at speeds up to 35 kilometers an hour (about 22 miles per hour).[45] Finally, during the annual cycle of the hive, drones are driven out of the colony to die every fall when the mating season ends and their usefulness to the community ends as well.

The life of a drone is unlike any other honeybee in the colony. Every sunny afternoon, once a drone reaches sexual maturity at about 12 days of age, he will fly from his hive looking for action. In ways that remain mysterious, he will find his way to one of the traditional honeybee mating areas ("Drone congregation areas") within a few miles of his home and will fly about this aerial pickup spot, waiting for a young queen to appear. If one does, he will zoom after her. If he manages to outrace his rivals and contact the queen, he will inseminate her in flight, 10 to 20 meters (30 to 60 feet) up in the sky.[46] The drone will then die following this event; his purpose as part of the hive has ended.

An architectural marvel

Beyond the social network of bees, the architectural marvel of the comb needs to be highlighted. All bees secrete wax for construction and routinely use the wax as a cap over the cell with honey for long-term storage.

*Comb*

So how is this comb created? Worker bees secrete beeswax in thin flakes from specialized glands in their abdomens. The workers construct comb using their jaws and the spines on their legs to shape and manipulate the wax flakes. It takes about a half-million flakes to make up a pound of wax.[47]

*Comb Design*

When completed, the comb consists of thousands of adjacent hexagonal cells, precisely engineered to remarkable precision, within 0.05 mm tolerance in diameter and about the thickness of a human hair in cell wall thickness. Hexagons evolved as the shape of cells for a good reason; it's the optimal design that leaves no empty spaces between cells while also packing in the greatest number of cells per unit area.[48]

## A Eusocial Insect

As one of the few eusocial insects (any colonial animal species that lives in multigenerational family groups in which most individuals cooperate to aid relatively few reproductive group member), honeybees show an advanced level of social organization in communication. Most animal communication is defensive or manipulative. Honeybees, like most other eusocial insects, communicate proactively, resulting in the interaction necessary to maintain such large groups.

After stroking each other's antennae for a few seconds to a minute or longer while also extending their long mouthparts, licking each other's tongues, heads, and legs and exchanging food, they also rub their front legs on each other's body parts. At other times we see bees vibrating while covering other bees or standing alone on the comb. Some buzzing is choreographed, with a pulsating bee circling in a round pattern or repeatedly turning in a distinctive figure-eight configuration.[49] We'll discussion communication in chapter 6.

*Communication*

What are the bees communicating? Most often, the bees communicate the queen's health, the hive's well-being, the location of flowers and plants, and when to swarm. Honeybees reproduce by swarming, in which the old queen and many of the workers leave the hive and form a cluster on a nearby tree branch or another overhang. The swarm sends out scouts, approximately 100, to search for a new nest site, which the swarm must agree on and move to within a few days if it is to survive and initiate the new colony.[50] The decision-making process of when, where, and how 60-70% of the bees choose a new home is enthralling. We'll discuss decision-making in Chapter 5.

*Relevance of a Colony Collapse Disorder (CCD)*

"During the winter of 2006-2007, some beekeepers began to report unusually high losses of 30-90% of their hives," according to EPA.[51] As many as 50% of all affected colonies demonstrated symptoms inconsistent with any known causes of honeybee death. What was found?

- Sudden loss of a colony's worker bee population with very few dead bees found near the colony.
- The queen and brood (young) remained, and the colonies had relatively abundant honey and pollen reserves.

But hives cannot sustain themselves without worker bees and eventually die. According to the EPA, "This combination of events resulting in the loss of a bee colony has been called Colony Collapse Disorder."[52]

Though agricultural records from more than a century ago note occasional bee "disappearances" and "dwindling" colonies in some years, it is uncertain whether the colonies had the same combination of factors associated with CCD.[53]

Why does any of this matter? Underlying every human reflection is our compelling need to understand who we are and why we are here, to consider whether the universe around us has meaning. Without a *mimetic* view (requiring humans to discover meaning that is already present in the world we inhabit) we won't. Situations in which the collapse of beehive colonies in 2006-2007 pass by as just another bizarre news article which will have little to no effect on our daily lives. Or will it? Since 65% of the advanced plants currently being cultivated require honeybees for pollination, we may need to pay attention.

Many theories about the cause of CCD exist, but the researchers who are leading the effort, as mentioned by EPA, are now focusing on these factors:

- Increased losses due to the invasive varroa mite (a pest of honeybees).
- New or emerging diseases such as the Israeli Acute Paralysis virus and the gut parasite Nosema.
- Pesticide poisoning through exposure to pesticides applied to crops or for in-hive insect or mite control.
- Stress bees experience due to management practices such as transportation to multiple locations across the country for providing pollination services.
- Changes to the habitat where bees forage.
- Inadequate forage or poor nutrition.
- Potential immune-suppressing stress on bees caused by one or more factors identified above.

## Summary

Consider the honeybee as a natural marvel. No wonder so many regions use the insect as a model for industrious activity. We should learn much from this little insect.

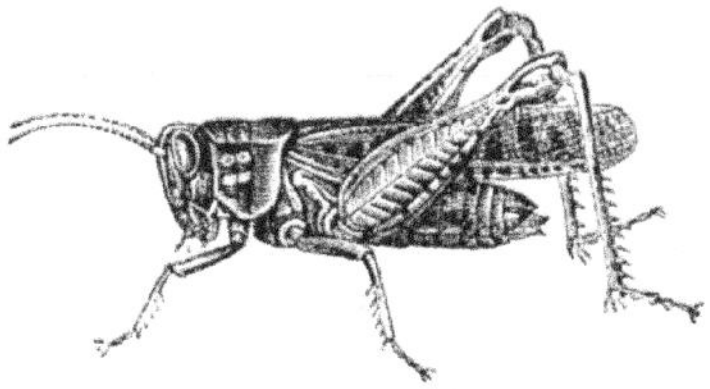

*Locusts have no king, yet they advance in ranks.*

– **Proverbs 30:27**

# Chapter 2: Locusts

My wife loves to garden. The front of our house looks somewhat like Tolkien's Rivendell without the waterfalls. Flowers, trees, and something in bloom almost all year long. Like Rivendell, the amount of foliage is almost overwhelming to the senses. Colors and fragrances seem to change daily.

She started her effort to plant flowers in our neighborhood before realizing how attractive these plants were to the nearby deer population. As a result, many flowers were beheaded shortly after they bloomed. The deer visited our home in the evening or early morning and took the edible plants away. My wife began to seek out and plant only deer resistant foliage. The deer now stay away and the garden continues to grow.

As she noticed my interest in bees and locust increasing, grasshoppers in the yard started taking on a new role. Grasshoppers are typically shy so seeing one is not common. Just recently, however, she captured a curious grasshopper climbing the stalk of a sunflower to look at her. They looked at each other for a while and the picture here is the curious grasshopper getting closer to my wife. Surprising how such a harmless little insect could be, if in the right situation, a scourge to the world of Biblical proportions.

---

Whereas honeybees are listed as one of the most helpful of all insects on the planet, locusts are at the other end of the spectrum. In fact, most people don't say the word locust without the word plague close behind. All locusts are grasshoppers, but not all grasshoppers are locusts. The study of this most feared insect is *acridology*, and one who pursues such a study is an *acridologist*. This is not an old term. It was first used to define the study of migratory grasshoppers or locusts less than 100 years ago. *Acrida* means locust—but it is also used as *acrid* which means having a strong irritating taste or smell. Perhaps the word choice is very appropriate.

## Dr. Jekyll and Mr. Hyde

If all locust are grasshoppers but not all grasshoppers are locust, how does this transformation occur? Most of society is familiar with Dr. Jekyll and Mr. Hyde and how a secret transformational potion changed one into the other. The difference is something like that. Locusts differ from grasshoppers because they have the ability to drastically change their behavior and habits as well as to migrate over large distances as a group.[54] Based on certain environmental conditions, a grasshopper—with the ability to become a locust—will change its behavior, color, body characteristics, and brain size. It's a scary thought. The solitary grasshopper (Dr. Jekyll), sitting quietly by itself in someone's backyard munching on lettuce, can become a gregarious, eating machine.

History of locusts

In the early part of the 19th century, *An Introduction to Entomology* included the following summary of this amazing insect.

> To look at a 'locust' in a cabinet of insects, you would not, at first sight, deem it capable of being the source of so much evil to mankind as stands on record against it. "This is, but a small creature," you would say, "and the mischief which it causes cannot be far beyond the proportion of its bulk. The locusts so celebrated in history must surely be of the Indian kind mentioned by Pliny, which was three feet long, with legs so strong that the women used them as saws. I see indeed some resemblance to the horse's head, but where are the eyes of the elephant, the neck of the bull, the horns of the stag, the chest of the lion, the belly of the scorpion, the wings of the eagle, the thighs of the camel, the legs of the ostrich, and the tail of the serpent, all of which the Arabians mention as attributes of this widely dreaded insect destroyer; but of which in the insect before me I discern little or no likeness?" Yet, although this animal is not very tremendous for its size nor very terrific in its appearance, it is the very same whose ravages have been the theme of naturalists and historians in all ages, and upon a close examination, you will find it to be peculiarly fitted and furnished for the execution of its office. It is armed with two pair of very strong jaws, the upper terminating in short and the lower in long teeth, by which it can both lacerate and

> grind its food – its stomach is of extraordinary capacity and power – its hind legs enable it to leap to a considerable distance, and its ample vans are calculated to catch the wind as sails, and so to carry it sometimes over the sea; and although a single individual can effect little evil, yet when the entire surface of a country is covered by them, and everyone makes bare the spot on which it stands, the mischief produced may be as infinite as their numbers. [55]

The Bible reminds us early of the ravage these insects did to Egypt:

The Eighth Plague of the Exodus

Then the Lord said to Moses, "Stretch out your hand over the land of Egypt for the locusts, so that they may come upon the land of Egypt and eat every plant in the land, all that the hail has left." So Moses stretched out his staff over the land of Egypt, and the Lord brought an east wind upon the land all that day and all that night. "When it was morning, the east wind had brought the locusts."[15] They covered the face of the whole land so that the land was darkened, and they ate all the plants in the land and all the fruit of the trees that the hail had left. "Not a green thing remained, neither tree nor plant of the field, through all the land of Egypt."[56] Rather overwhelming. They ate everything.

Yet, if you look at a grasshopper in your backyard, it doesn't appear that ferocious. Why all the excessive descriptions? What has this insect done to deserve such treatment? Furthermore, what does a locust really look like?

Let's stay with biblical sources. In the book of *Revelations:* "And the shapes of the locusts were like unto horses prepared unto battle, and on their heads were as it were crowns like gold, and their face were as the faces of men." In the book of Joel, "their appearance is like the appearance of horses, and like war horses so shall they run." Again in Joel, describing the locust and their mouth as "powerful and without number; its teeth are lion's teeth, and it has the fangs of a lioness."[57]

Are these descriptions of Locust correct? Yes. All these biblical descriptions would be approved by an entomologist. Looked at from underneath, the mouth parts of the *Schistocerca gregaria*, the desert locust, do bear a strong resemblance to a lion's muzzle surmounted by nose and eyes. The upper part of the head,

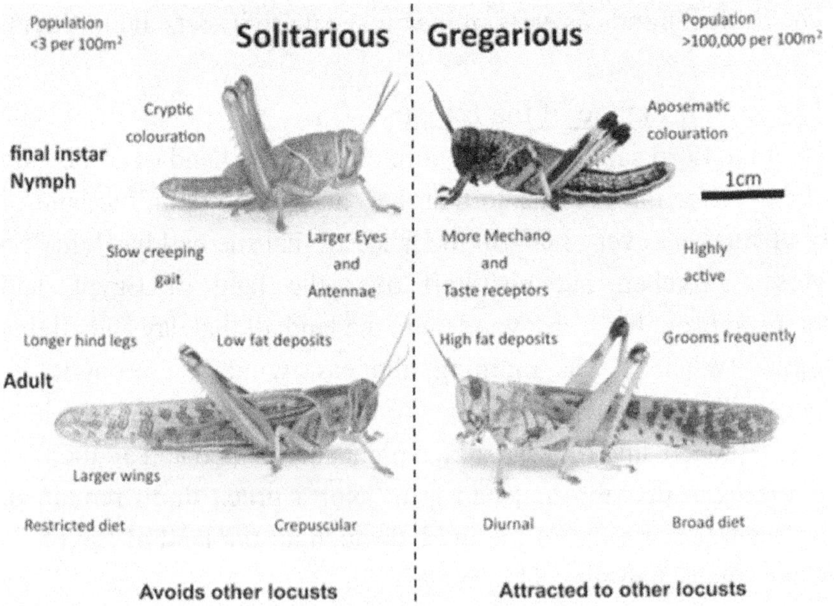


when viewed from the same angle, looks like a crown. Yellow adds to the crown of the mature gregarious locusts. The muzzle looks like a human face, while the head, seen sideways, is that of a stylized horse. German peasants actually call a locust a Hay Horse—*heupferd.*[58]

The Desert Locust

Let's look particularly at the history of desert locusts. The desert locust species seems to devastate Africa frequently. In the first century a law in Cyrenaica (the eastern region of Libya) mentioned by Pliny (a historian in the first century), drafted citizens to destroy the locusts in three different states, three times in the year—first their eggs, then their young, and lastly the mature insect. Why such a law? Because entire cities were devastated and the war against desert locust required drafting all citizens in the fight.

An early historian named Orosius (a student of Augustine circa A.D. 415) documented a time when Africa was infested by these animals such that, having devoured every green thing, after flying off to sea they were drowned, and being cast upon the shore they emitted a stench greater than could have been produced by the carcasses of 100,000 men. In the same history book, Orosius mentions a plague that arose in Masanissa (North Africa) from the same cause, which destroyed no less than 800,000 persons and many more in the territories bordering the sea.[59]

So, if the stench of these insects could produce an odor like the rotting carcasses of 100,000 people, how large is a desert locust swarm? A square kilometer of a well-packed swarm observed in East Africa contained 40 million locusts. Some estimates state that a swarm could contain 150 million locusts per square kilometer. A large swarm, 100 square kilometers could contain 10 billion locusts.[60] More locusts in less than 40 square miles than we have humans on the earth! This is a social (gregarious) insect!

Of course, this is a frightening background story but behavior change? Dr. Jekyll and Mr. Hyde? How is this possible? As locusts increase in number and become more crowded, they change their behavior from acting as an individual (solitarious if you're an entomologist) insect to acting as part of a group (gregarious). The appearance of the locust also changes—solitary adults are brown, whereas gregarious adults are pink (immature) and yellow (mature). Up until 1921, it was thought that the locust was a different species of grasshopper that appeared out of nowhere![61] How did someone uncover this Dr. Jekyll to Mr. Hyde transformation?

## Boris Uvarov (1886-1970)

The first step in the solution comes from an intriguing Russian entomologist who discovered this transformation approximately 100 years ago. His name is Boris Uvarov. He postulated, after observing what appeared to be the transformation of a species of grasshoppers into a locust, that locusts must have phases. A phase change occurs when a single species transforms from what appears to be one species to another and back again based on ecological triggers and physiological mechanisms.[62]

What did Dr. Uvarov discover? On the Western Steppes of Russia, Uvarov noted that *danica* (solitary locusts) would leap out of the way and abandon the gathering of *migratoria* (gregarious locusts) of the migratory locust group. Initially, he assumed the two were different species because of 1) color differences, 2) shape of *pronotum* (the hard shieldlike covering of the fore part of the thorax behind the neck), and 3) the ratio of the length of the *elytron* (wing case) to the hind femur. Furthermore, *danica* did its best to avoid the other group.[63] But after following the migratory locust, he discovered the *danica* again in and around the swarm. The *danica* were passive, reclusive, and solitary. The *migratoria* were social, outgoing, and desirous of working together. Could these insects be the same?

In his 1921 publication, having discovered the connection between the two-phase insect, Uvarov then revealed the secret to fighting locusts (*Bulletin of Entomological Research* September 1921):

> The direct cause of [this] ignorance is that injurious insects, and locusts especially, are studied only in the years of maximum development, and nobody cares about them in the minimum years when the clue to the whole locust problem is likely to be found.[64]

Uvarov's insights into the proper identification of *locust migratoria* were eventually recognized as "the greatest contribution ever made by a taxonomist in the solution of a major economic problem."[65] His revelation of a Dr. Jekyll and Mr. Hyde transformation is startling. So, what is a phase change?

During normal years, the locusts persist in their solitary phase, generally avoiding one another to exploit the available resources more evenly. In moist years, their populations begin to increase within their breeding areas. This increase may be sufficient to cause crowding in some habitats, such as deserts, where food is normally scarce. [66]

There are thought to be between 10-15 different acridids that can transform from a grasshopper to a locust. The more well-known are the migratory locust, which has a wider range than any other acridid. It is found in grasslands throughout Africa, most of Eurasia south of the taiga, the East Indies, tropical Australia, and New Zealand. The desert locust inhabits dry grasslands and deserts from Africa to Punjab and can fly upward to about 1,500 meters (5,000 feet) in huge towers of individuals. The smaller Italian and Moroccan locusts cause extensive plant damage in the Mediterranean area, with the Moroccan locusts being found as far east as Turkestan. In South Africa, the brown and red locusts are extremely destructive. The chief migratory species in Central and South America is the South American locust. The nonmigratory South American locust may be a solitary phase of this genus. The Rocky Mountain locust and the migratory grasshopper destroyed many prairie farms in Canada and the United States in the 1870s. Many other species occasionally increase sufficiently in numbers to be called plagues.[67] Thus the phase change from solitary to gregarious is on a sliding scale. All grasshoppers have a tendency to change but 10-15 *acridids,* being the most infamous, have gained the name locust for their extensive transformation and impacts on society.[68]

The following diagram depicts the transformation and is used to train responders in the desert locust region of North Africa.

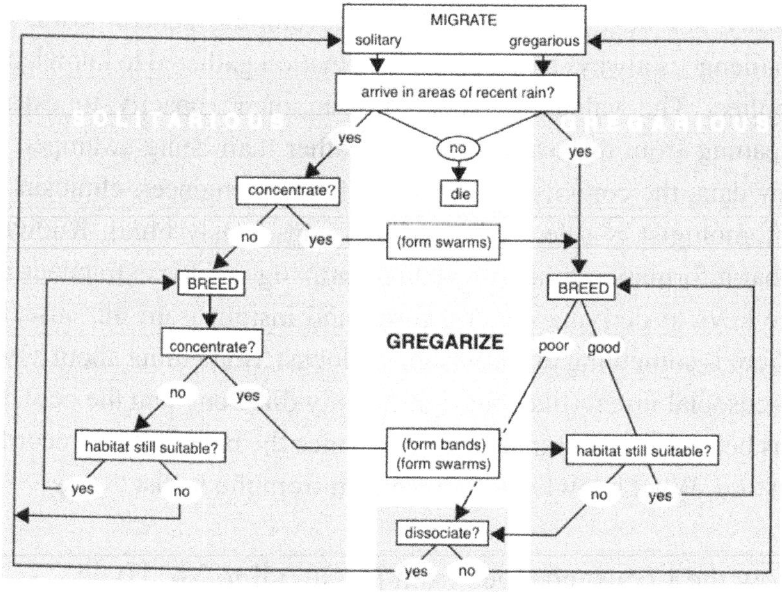


## Why is the African coast and the Middle East so susceptible to locust swarms?

In the invasion area of the desert locust (*Schistocerca gregaria*), the infliction is greater because the humans living there are poorer, and their livelihoods are at best marginal. The land is poor, they work hard, and everything grown by hand can be destroyed in a few hours. The most troubled area seems to be in the most danger. The Nile Valley is particularly susceptible with the rich agricultural region susceptible to locust plagues. The area impacted by the desert locust is twice the size of China and supports 500 million people.[69] Though it may appear sparsely populated, a locust plague can impact a large section of the human population.

Are you drowning in information yet? We have more data than we can possibly use regarding nearly every matter of society, economics, politics, and now insects. Science has contributed to this tsunami of facts and figures, but the tidal wave of minutiae is hardly unique to our time. When developing general laws, the challenge is always to move the information gathered to knowledge applied. The value of experts lies in their capacity to extract meaning from the reams of facts. Rather than being swamped by raw data, the connoisseur, artist, craftsman, engineer, clinician, or entomologist is selectively and self-consciously blind. Knowing what information to ignore and recognizing what is extraneous are the keys to deriving pattern, form, and insight from this insect.[70] There is something to learn from the locust. Something about a type of eusocial insect (like, but significantly different than the bee) that has been a bane to human existence since the beginning of recorded history. What knowledge can we gain from the locust?

## How the Locust changes from Dr. Jekyll to Mr. Hyde

[71]To move from the lonely, solitary grasshopper, intent on a life of separation, to the social animal known as the locust routinely takes

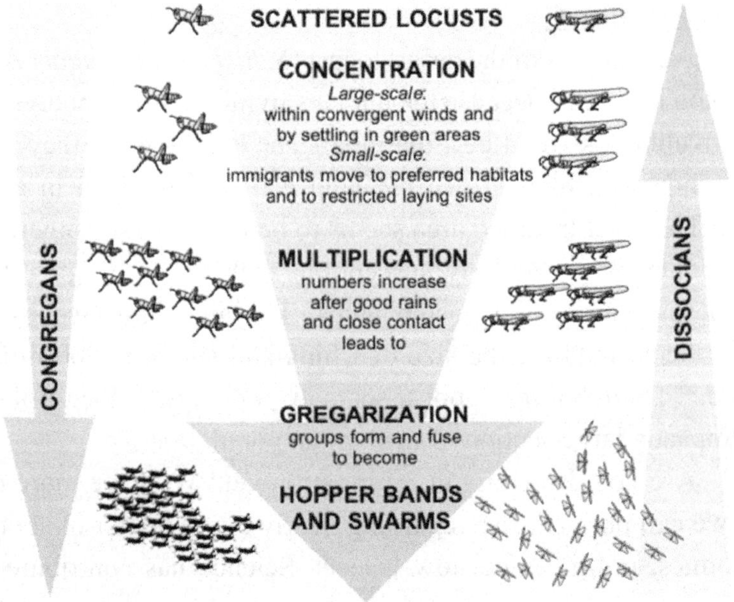

four generations or cycles. Each cycle steps the locust through the behavioral and physical changes resulting in a plague.

*Locust life cycle 1*

First, there is an unintentional gathering of the solitary locust. Since the insects are always in search of food, a super-abundance of food following desert rains in a normally dry region results in an increase in the locust population. The infrequent rains then stop, and the greenery is either consumed or dies. The solitary locusts are then forced together during these dry spells when minimal areas of green plants in the desert exist.[72]

*Locust life cycle 2*

Second, a sudden transformation of these solitary locusts occurs. Within hours of crowding, a boost of serotonin in the central nervous system spurs behavioral changes such as rapid movements, sociability, increased self-grooming, and a more varied appetite.

*Locust life cycle 3*

Third, a breeding and feeding frenzy ensues. In a perfect storm, returning rains provide moist planting ground for future generations of gregarious locusts. New generations, with the gregarious tendencies of their parents, form groups that forage in unison. Once their wings grow, the herds become flying swarms in search of more food and habitat. Next is a toxic new look—swarming locusts turn bright yellow—a warning to birds, reptiles, and other predators that they may be unappetizing or even poisonous.

*Locust life cycle 4*

Finally, we have a flying apocalypse. The mature locust joins with other swarms and grows in number. A major plague begins when swarms, often traveling 30 to 60 miles a day, develop over multiple regions. Without intervention, plagues can last years until a natural die-off occurs.

## The strength and powers of Mr. Hyde

As stated earlier, the solitary grasshopper now has an altered anatomy, including physical changes such as a larger brain and shorter legs. The brain boost, especially the midbrain, grows larger, perhaps for more complex information processing because of the increased number of other flying locusts in proximity. The locust also has increased endurance, and the offspring of gregarious locusts develop smaller femurs. Their hops are shorter but use less energy to cover the same distance.[73] Thus, in three generations (four cycles), what appears to be a harmless grasshopper has become a member of the most feared of all insects—a locust plague. Mr. Hyde in all his terror.

## Why don't all grasshoppers become locusts?

This is not an easy relationship to explain, so let me do so with an analogy. Locusts are to grasshoppers as athletes are to humans. Nobody would doubt that a professional basketball player is an athlete (think for example of LeBron James), but what about a professional golfer or bowler (think of John Daly)? The softball player? Thus, being an athlete is like being a locust—it is a matter of degree. Some creatures exhibit weak tendencies toward aggregating as nymphs or adults, and others manifest the full-blown locust syndrome—the Rocky Mountain locust and the desert locust being classic cases at the extreme end of the locust spectrum. At the other end of the spectrum are the various garden-variety grasshoppers that placidly masticate our lettuce patches and bask in abandoned lots.[74]

One can also see that the ecological environment of the desert or the plains of middle America prior to the 18$^{th}$ century was perfect for the locust's development. Just like a prize athlete thrives on the field of competition best suited to their skills, so do the desert locust. In fact, the desert locust seems to thrive in the very worst of environments. When conditions are favorable for reproduction, locust numbers increase, and when they are not, numbers decrease either by natural mortality or through migration.[75]

## The five/six instars

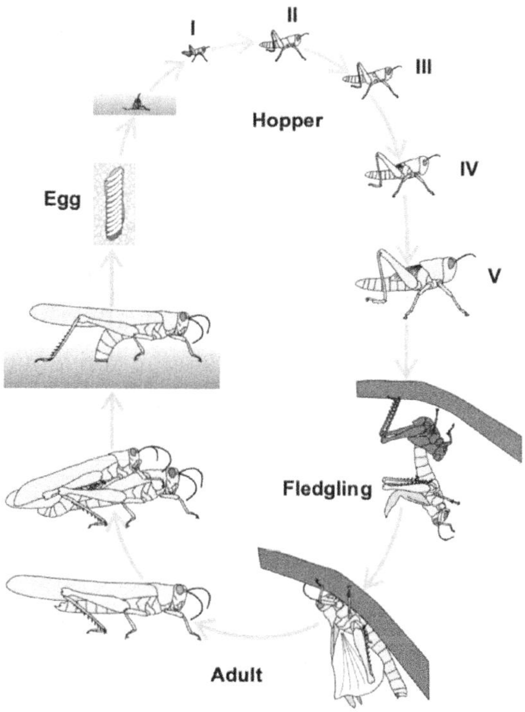


[76]So how do locusts develop? Thirty to eighty eggs are deposited as a single pod by the female in the soil. The embryos mature through the summer and then hibernate (all underground)—a physiological state called diapause in insects—for the winter. The following spring, as the ground begins to warm, the soil appears to boil as the nymphs hatch and crawl to the surface. The nymphs will molt or move through five to six instars (growth periods where it becomes necessary for the

insect to shed its exoskeleton to continue that growth) while marching in bands across the land. The molt would result in the shedding of their old outer cuticle while growing through each instar. On the final molt, the stubby wing buds are replaced with fully developed wings, the reproductive system becomes functional, and the adult locust is ready to swarm.[77] This is the normal developmental route for all locusts despite the phase (solitary or gregarious) that they are in.

*What internal environmental changes occur?*

The gregarious phase is triggered by the insects' endocrine system (increase in serotonin), which is precipitated both by chemicals released from the crowded insects via their accumulating excrement and by the frequent disturbance of tiny hairs on the hind legs of the creatures.[78] Thus the environmental change results in the locust crowding and the odor and contact between the locust prompting the transformation. Simply stated, the gentle, solitary 'Dr. Jekyll' grasshopper with a major serotonin rush has now become a marching and eventually flying army of 'Mr. Hyde' locusts prepared to devastate the economic well-being of a region in a matter of hours. A plague has begun.

*Is a plague an unnatural extreme?*

Extremes don't exist in nature for long. Nature is seldom so absolutely divisive. Extremes are, overall, abhorrent to it. Even in the worst 'years of the locust', there is likely to be some time or place at which the insect seems poised uncertainly between two opposing poles, one toward solitary living and the other toward swarming. Like Dr. Jekyll aware of his evil twin, the locust longs for the quiet life of living, raising a family, and playing with the grandkids (perhaps). Yet a rush for the excitement of change and adventure lurks in the shadows. The flashy new look (yellow and black) and the never-ending party calls like the siren song of disaster in Homer's *Odyssey* to lure the solitary locust to move closer and closer. Long periods of time make the transformation from one phase to another more distinct and provide the watchful public with an appreciation of a devastating sequence which impacts the entire cycle of life.

Such extremes of behavior within one species must seem almost baffling to humans—indeed, only the human species can rival it. Nor does the nomenclature employed to describe this insect clarify the picture—for how can it be solitary and then gregarious? The insect seems to live two separate lives. The city-dweller and the country squire. Migration to groups, and then movement away from groups. The change appears not so much based on location to each other (thousands or even hundreds of thousands of locusts are in the same wadi—a valley, ravine, or channel that is dry except in the rainy season—but all solitary[79]) but a grouping, a serotonin rush, the odor of excrement, and the disturbance of tiny hairs on the hind legs changes everything—truly an unusual insect.

Of course, locust swarms continue even today. The second largest swarm ever reported occurred in Kenya in1954 when fewer than a hundred square miles were covered by an estimated tens of billions of desert locusts. Note that was the second largest swarm. The largest swarm occurred in the U.S. approximately 150 years ago. No insect outbreak has ever approached the magnitude of the Rocky Mountain locust reported in 1875 which may have contained as many as 3.5 trillion insects![80] (See Appendix A for a synopsis of the 1875 Albert's Swarm).

<u>*The Impact of a Desert Locust Swarm*</u>

How does a society fight such a pest – especially the desert locust? During peaceful periods (known as recessions), desert locusts are usually restricted to the semi-arid and arid deserts of Africa, the Near East, and South-West Asia that receive less than 200 mm of rain annually.[81] Without the input of rain, the population remains level. But when the rain comes, the change begins. The appetite of the desert locust is voracious.

A desert locust adult can consume roughly its own weight (which is about two grams) of fresh food daily. A 1 $km^2$ size swarm contains approximately 40 million locusts, which can eat the same amount of food in one day as 35,000 people. This is based on a person eating an average of 2.3 kg of food per day, according to the USDA.[82]

To put this impact into perspective, in one day a swarm:

- The size of Paris eats the same amount of food as half the population of France.
- The size of New York City eats the same as everyone in New York and California.
- The size of Rome eats the same as the population of Kenya.
- The size of Sydney (Australia), eats the same amount of food as Australia eats in 1.5 hours.[83]

So, what do you do? First, to fight such a pest, one needs to discover where they breed. Between 1958 and 1963, immensely valuable work in this respect was accomplished by George Popov who established for the first time, in detail, the kind of habitats favorable to desert locust breeding and where, in general, they were to be found. Unhappily, his FAO/UNDP Ecological Survey also confirmed beyond a doubt that the insect has no single defined breeding areas—a swarm can break out anywhere.[84]

*The fight against the Desert Locust*

With the potential for the desert locust to impact 60 different countries, the first layer of defense and control are the responsibility of the Ministry of Agriculture in locust-affected countries and the operations undertaken by national locust units. There are also several regional locust organizations that assist with defense and control operations. During times of outbreaks and plagues that cross national boundaries, the Food and Agriculture Organization of the United Nations joins the fight.[85]

Some may ask is the problem really that big? Why does the U.N. need to be involved? Well, the answer to that question concerns the speed at which the desert locusts can migrate. Desert locusts usually fly with the wind at a speed of about 16-19 km/hr. Swarms can travel up to a distance of 5-130 km or more in a day. Locusts regularly cross the Red Sea, 300 km with not much effort. In the past, there have been some spectacular and very long-distance swarm migrations. For example, elements of some locust swarms traveled from North-West Africa to the British Isles in 1954 and from West Africa to the Caribbean, 5,000 km in about ten days in 1988.[86]

## Summary

*Stages*

A Desert Locust lives approximately three to five months, although this is extremely variable and depends mostly on weather and ecological conditions. The life cycle comprises three stages: egg; hopper; and adult. Eggs hatch in about two weeks (the range is 10-65 days), hoppers develop in five to six stages (molts or instars) over a period of about 30-40 days, and adults mature in about three weeks to nine months but more frequently from two to four months.[87]

The desert locust has five molts, or the changing of its exoskeleton, before achieving adulthood. The young locust hopper is, in effect, exchanging its suit of armor to accommodate its growing body.[88] The insect form developed in between two subsequent molts is termed an instar. Between the first and second instars, there is little change except to the eye of the expert, who notes the intensified coloration. With the third instar, the wing buds can be seen, a small wing overlying a larger one. In the next stage, the wings are inverted from a lateral to a dorsal position and become well-marked and distinct. The position is reversed once again at the last molt. All this takes from three weeks to two months or more, depending on local temperatures. As the last instar ends, the hopper is ready to become a flying locust.[89]

*Locust Controls*

At present, the primary method of controlling desert locust swarms and hopper bands is with organophosphates applied in small, concentrated doses (referred to as ultra-low volume (ULV) formulation) by vehicle-mounted and aerial sprayers and, to a lesser extent, by knapsack and hand-held sprayers.[90] However, novel pheromone research is having its impact as well as the cannibalistic tendencies of the locust—more about that in chapter 7 addressing balance.

Perhaps you're saying—what a creature! A devouring insect that can number in the billions and march or fly into an area from hundreds of miles away, capable of consuming the entire landscape and all edible food in hours. Our only response is hitting the horde with organophosphates or modern pheromone techniques. How frequently does this occur? There is no evidence that desert locust plagues occur after a specific number of years. Instead, plagues develop intermittently. During the last century, desert locust plagues occurred in 1926-1934, 1940-1948, 1949-1963, 1967-1969 and 1986-1989.[91] For the most part, the ravages seen in the past are occurring less and less frequently. See the table from the UN FAO Office below.

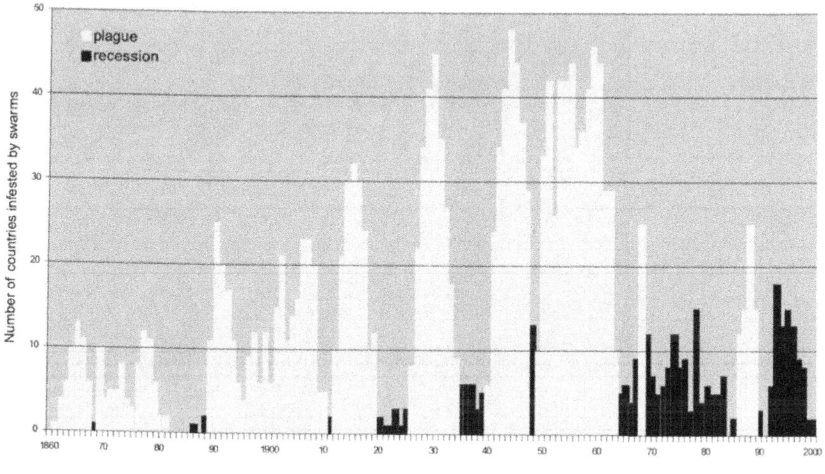


Figure 18. Plague and recession periods of the Desert Locust, January 1860 to December 2000.

92

The locust is arguably the most hated of all insects on the planet. What can we learn from it? Perhaps some lessons about handling stress.

 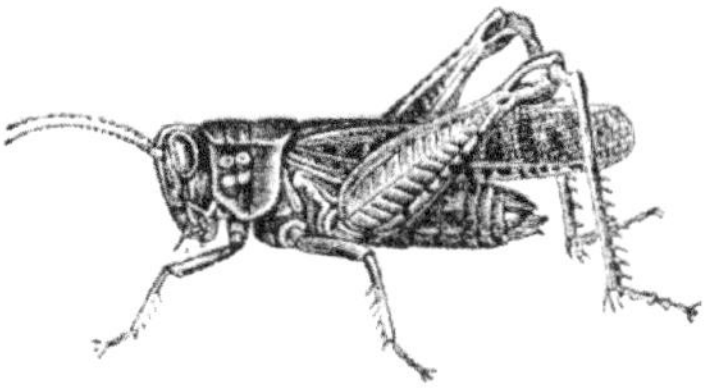

*It is not stress that kills us, it is our reaction to it.*

– **Hans Selye**

# Chapter 3: Stress

When my youngest son was three years old, he stopped walking and went back to crawling. He had been toddling and running across the house and yard like any happy, carefree youngster, and then he stopped. He would crawl out of his room, down the stairs and sit, grabbing at his ankles and crying but not walking.

The early doctor visits were not helpful, but eventually we found the right specialist who was able to detect osteomyelitis through an MRI. Osteomyelitis is a serious infection of the bone that can be either acute or chronic. It is an inflammatory process involving the bone and its structures. Yet during the MRI, the doctor also announced that my son had Acute Lymphoblastic Leukemia (ALL) and would need to begin Chemotherapy as soon as possible.

My wife and I were in shock. The University of North Carolina Hospital at Chapel Hill quickly accepted my son and the procedures began. The first week was the most stressful. The new chapter in our life seemed more like a nightmare. How could it get worse? Yet the social worker opened our eyes to the fact that it could get worse, much worse.

She appeared concerned about our condition and asked if we were handling the stress.

"Yes," we told her. She wanted to make sure we were ready for the ordeal, and we mumbled that our son would need a lot of support.

She looked at us despairingly and said, "Over 60% of families that have children fighting leukemia will go bankrupt, and 70% of these marriages will end in divorce."

We looked back in horror, "Really?"

"Yes. I advise you to prepare for a family-wide onslaught on what you once called normal."

We almost collapsed—the stress was on.

Over the next three years, we held tenaciously to our son, our family, and each other. We sold everything that was a luxury and watched with tears and prayers as the "chemo cocktail" was injected into that little body day after day, week after week, month after month, and year after year. Yet we didn't weaken as a family. In fact, we became stronger.

---

The body is a finely tuned instrument that does not like surprises. Any sudden change in stimuli which affects the body, or the reordering of important routines to which the body has become accustomed, can cause stress, and throw one's whole physical being into turmoil. Your nervous system releases a flood of stress hormones, including adrenaline and cortisol, which rouse the body for emergency action. Your heart pounds faster, muscles tighten, blood pressure rises, breath quickens, and your senses become sharper. A decision of what to do next is before you. Perspiration begins to build on your forehead, even in a cool climate. Your palms begin to sweat. Should you fight or run? You have a decision to make. What will it be?

Psychologists have developed ways to gauge the impact of these surprises. In fact, many are familiar with the Social Readjustment Scale (see Appendix B). To use this tool, you check off variation in your normal routine, take the rated impact number, sum the total and compare it to the rating at the end of the test. Numbers are associated with being laid off, moving, death of a spouse, change in the route traveled to work, etc. Since being healthy is the optimum state one wants to achieve, being sick, and in this case, stressed, is the state one most wants to avoid (if possible).[93] Of course, you get some outrageously high numbers if you take the chart and use it during wartime, pandemics, or major weather conditions (tornado, hurricane, or flash floods). Numbers would make one pause and wonder how anyone could successfully manage to live through these accumulative events.

For example, the stress caused by the pandemic significantly impacted teenagers and young adults, according to a study completed by the research group Plos One. "Strikingly, younger adults showed disrupted maturity in that they increased in neuroticism and declined in agreeableness and conscientiousness."[94] The result? A disagreeable, nervous, and self-centered group from the stress of COVID-19. Some may think humans could never learn anything about managing stress from observing a group of bees or locusts—they can't possibly confront the same stressors managed by human societies. Well, perhaps not the same stressors, but stress, nonetheless. And situations that one can learn from these insect communities may provide help with our own stress management.

# Honeybees

## The Story of the Colony Collapse Disorder

Bees pollinate about one-third of all crops, including fruits, berries, nuts, vegetables, and the forage crops like alfalfa, which are used for hay to feed livestock. In fact, of the 100 top edible crops that flower, 80 are pollinated by honeybees. There aren't many wild and native pollinators left in large numbers. Without honeybees, we would be hard-pressed to feed ourselves.[95]

Colony collapse disorder (CCD) is an abnormal phenomenon that occurs when the majority of worker bees in a honeybee colony disappear, leaving behind a queen, plenty of food, and a few nurse bees to care for the remaining immature bees. While such disappearances have occurred sporadically throughout the history of apiculture and have been known by various names, the syndrome was renamed colony collapse disorder in early 2007 in conjunction with a drastic rise in reports of disappearances of western honeybee (*Apis mellifera*) colonies in North America (see beeinformed.org).

The first report classified as CCD was in mid-November 2006 by a Pennsylvania beekeeper overwintering in Florida. By February 2007, large commercial migratory beekeepers wintering in California, Florida, Oklahoma, and Texas had reported heavy losses associated with CCD. Their reports of losses varied widely, ranging from 30% to 90% of managed bee colonies. In some cases, beekeepers reported losses of nearly all of their colonies, with surviving colonies so weakened that the beekeepers questioned whether the bees might not be able to pollinate or produce honey at all in 2007.[96]

A colony that has collapsed from CCD is generally characterized by all the following conditions occurring simultaneously. The presence of:

- The queen bee
- No dead honeybee bodies in the vicinity
- Capped brood in abandoned colonies. (Bees normally do not abandon a hive until the capped brood has all hatched)
- Healthy food stores
    - That other bees do not immediately rob
    - With significantly delayed attacks by hive pests such as wax moths and small hive beetles[97]

Almost like a mass suicide, 20-50,000 bees decided to fly off into the sunset, leave their community and food source, and die alone. Why?

<u>Did human society cause the collapse? Are we the source of stress?</u>

Some would say yes. USDA Research Scientist Jeff Pettis stated, “We believe that subtle interactions between pesticides and pathogens, could be a major contributor to increased mortality of honeybee colonies and other pollinator declines worldwide.”[98] For example, David Hawthorne and Galen Dively of the University of Maryland published a study with the telling title “Killing Them with Kindness”, reporting that the impact of miticides applied by beekeepers is more severe in the presence of antibiotics, which reduce the bees’ ability to detoxify and eliminate pesticides themselves.[99] In an attempt to make the hive healthy and to reduce mites and other pests from the hive, humans may have inadvertently weakened the natural immunities within a colony. Each cut is not that much of an issue, but death from a thousand cuts—could this be the reason for CCD?

Concerning our own health, antibiotics are very important. However, could our own weakened immune system, from numerous vaccinations be a concern for sustainability? Some pharmaceutical companies collect and process human blood plasma. The proteins in the plasma are the treasure trove of the eventual products developed. However, the plasma collected in areas of the country where the population is less immunized has unique and sometimes valuable protein characteristics. Individuals in these locations developed, through their own physical hardships and immune responses, proteins that others lack. Like the bees that have not been dosed with excessive antibiotics, does this make the human populations in these areas more sustainable? Perhaps.

Another theory, according to Wani Zhu, a graduate student at Penn State who used birth, fertility, life expectancy, and death data to predict population size and age structure as well as the impact of pesticide and disease synergies identified in previous research stated:

> These subtle effects can include slightly reduced nectar and pollen collection due to disoriented foragers failing to return to the hive, shortened life spans to workers who begin to forage a day or two earlier than they normally would, and a small increase in larval mortality, which result in a downstream reduction of the adult worker population. By destabilizing the age structure that characterizes a healthy colony, disrupting the well-functioning balance of eggs, larvae, pupae, and adults of various ages. What happens? Precocious foraging by adult worker bees is caused by some

> pesticides and diseases that interfere with a hormone that regulates the ages at which bees begin to forage. Early foraging removes younger bees from in-hive tasks such as feeding the young, resulting in higher brood mortality and other cascading effects in a downward spiral of colony decline.[100]

Zhu connects the stress of a shortened live span to causing the colony breakdown.

> It's as if most adult humans … were to wander off over a few weeks or months, leaving babies and toddlers to be tended by teenagers. Some teens would prematurely become responsible caregivers, but it wouldn't be long before our age-shifted human society would collapse into turmoil and dysfunction.[101]

The lesson concerning sustainability—disruption in the maturity of caregivers—could severely limit a societies ability to continue. Caring for the young and the old is important to sustainability.

Some believe that colony collapse disorder was inevitable, a disaster for bees and beekeepers alike and, for our food supply, one brought on by our best human intentions to manage the world around us. Each cut is insignificant on its own but deadly in the context of many other cuts. That's what's happened to bees. Myriad of afflictions, each controllable alone by the bees' natural resilience, finally crossed the line where synergy kicks in.[102] The stress of a thousand cuts resulted in 30-90% of hives calling it quits in 2006-2007.

## Lessons from CCD

Complex systems like bee colonies or human societies are marvelously functional when healthy, but their very complexity makes social organisms particularly vulnerable to sudden and epic breakdowns when overcome by too many challenges or too many stressors. Perhaps our human ingenuity is remarkably effective at solving individual problems, but our blind spot is our inability to recognize when interactions between issues might bring us down. Go back to the Social Readjustment Scale (Appendix B) mentioned in the introduction. Market limitations on toilet paper or basic food supplies seem to result in a short-term panic we laugh about months later; the occurrence though tends to upset the balance we believe is necessary to survive and sustain our communities. Yet bees are not known for collapsing under stress, and their basic reputation is one of hardiness and diligence. Similarly, certain human populations have shown this hardiness as well. Perhaps immune strengths, gained through difficulties, may provide a hardiness we need. If the world possesses intrinsic meaning, as understood through a *mimetic* view, challenges and stress are beneficial.

## Stress and the bee work ethic

Bees are symbols of an excellent work ethic. In her book *Bees in America*, Tammy Horn notes: "Bees fit in well in a place where Christian values merged with the Puritan work ethic to create a culture in which industry and efficiency equal financial reward and respect."[103] Muslims say that true believers are like bees. One Islamic scholar wrote, "Honey is a remedy for every illness, and the Qur'an is a remedy for all illness of the mind. Therefore, I recommend to you both remedies, the Qur'an and honey."[104] Even the Utah flag proudly emblazoned a honeybee at its center, and the states of Arkansas, Georgia, Kansas, Louisiana, Maine,

Mississippi, Missouri, Nebraska, New Jersey, North Carolina, West Virginia, and Wisconsin claim the honeybee as their state insect. The collapse of the honeybee community is the exception, not the norm. So, what drives honeybees' activities and their management of stress? Survival.

Stressed about survival.

Rob Page of the University of Arizona states that a hive's entire superorganism is based on the insect's diligence to produce the food necessary to survive. The worker is told by the larvae that they need to be fed. The other workers, based on the availability of what is in the hive, begin collecting how much pollen is available for feeding larvae. Every decision is based on food. The *Spirit of the Hive* (the name of his book) is one of needed consumption.[105] Stress impacts the completion of the key tool for sustainability – food production. Thus, the age difference will change if the hive is short on pollen, and more bees will be sent out to start foraging. If pollen is adequate, fewer bees will forage, and the age difference will change again.[106]

Similarly, Paul Ehrlich in 1968 published *The Population Bomb* with a similar but only partial premise – food is the critical input for human survival and society can do nothing about changing the amount of food. In the 1970s, the book promised, "hundreds of millions of people are going to starve to death." No matter what people do, "nothing can prevent a substantial increase in the world death rate."[107] Ehrlich's dire predictions of a "human colony collapse disorder" never came to pass and his promotion of population control, as opposed to increased food production—a lesson we observe from the bees—have damaged society in numerous ways. Population continues to increase, but the amount of food being produced per acre is increasing as well.

## Beehive chaos and stress from governmental disorders

The most basic definition of stress, according to the CDC, is a state of worry or mental tension caused by a difficult situation. For honeybees, the major stressor may not be the adequate availability of pollen, but something humans can easily see we share with bees—stable governance. When events prompt the tragedy of governmental breakdown—or even the change of your current supervisor to a new supervisor—the chaos of the peaceful order of a well-functioning society (or business) ends. Human history is more than replete with war and violence, considerably more so than the honeybee world, but even the placid world of the honeybee colony can descend into conflict if the queen suddenly dies. The workers sense her loss within minutes due to diminished queen odors in the hive and begin rearing a new queen.[108]

## Long live the Queen!

Without a queen, the hive will dwindle and die within a few months. As it declines, worker bees become highly aggressive toward each other and any unfortunate beekeepers who might open the nest realizes that all is not well in Beeville. Some of the workers develop their ovaries and lay eggs, which develop into male drones since they are not fertilized. Vicious fights break out between workers trying to become dominant as egg layers, which is a heartbreaking act of desperation because, dominant or not, their demise and that of the colony are inevitable without the rearing of new female workers (to be discussed in greater detail in Chapter 7). The most cooperative of societies, even that of honeybees, can collapse into disorder and violence given the right circumstances in a few weeks. Of the four hives that I oversee at home, one is suspiciously aggressive. For a short period, it almost descended into the stressful chaos mentioned above. I could hear angry bees and see the increased drone population. Fortunately, I found the queen. She was young and recently returned from her mating flight, with enough area in the brood box to reestablish the necessary balance. Fortunately, the hive survived. Prior to the swarming of the hive (which caused the stress), the leaving queen had successfully laid enough queen cells to get through the transition.

*The Queen's line of succession*

The queen is the stabilizing influence of the hive.

First, the longevity of her command. The queen will routinely live for 1-4 years, reducing the stress of a change in structure. The other bees live only a fraction of this time (4-6 weeks) so her just being there has a calming impact.

Next, her communication of the health and purpose of her command. The condition of the queen (her well-being, presence, age) is perhaps the most important news being communicated throughout the hive – we lose her (they appear to communicate) – and, as a hive, we fail to respond quickly, the colony will cease to exist. The communication (her pheromones) tells the story to all other bees and must prompt quick action if the hive is going to survive.

Finally, multiple future queens are developed when the original queen dies or prepares to leave. When the hive is going to swarm and create a second hive, the existing hive prepares numerous queen cups (a special cell for future queen growth). The replacement queen is not routinely a single bee or heir to the throne. Multiple larvae are given royal jelly (the necessary substance that changes the DNA of the worker bee into a queen), and then contend, after leaving the cell, for this most important role of queen in the hive.

Leadership in human societies is not so straight forward. If only "royal jelly" could be fed to a group of select humans who could then live for centuries and, with the wisdom of the ages gained from the "royal jelly" and the experience of all those years, direct our social and political systems. Of course, humans require both experience and education to successfully take and manage large organizations in stressful situations. Yet this combination of education and experience was not always seen as necessary. In the United States, the training of military officers was once questioned by the famous Davy Crockett.

Davy Crockett

In February of 1830, a young Congressman from Tennessee believed that experience was more important for leadership than education. As a result, he issued a proclamation to close the United States Military Academy at West Point and have all officers chosen from the ranks. The importance of a military education he believed, was secondary to courage, audacity, and resolve.

The resolution did not pass but the lingering suspicion of what is needed to develop leaders to manage the chaos of war continued. The Mexican American War from 1846 to 1848 put much of that to rest with ~80% of all line-officers in the conflict originating from the Academy. General Winfield Scott, the commanding general of the U.S. forces during the conflict stated clearly:

> I give it as my fixed opinion, that but for our graduated cadets (from West Point), the war between the United States and Mexico might, and probably would have lasted some four or five years, with, in its first half, more defeats than victories falling to our share; whereas, in less than two

> campaigns, we conquered a great country and a peace without the loss of a single battle or skirmish.[109]

Developing leaders doesn't require royal jelly, but a necessary combination of both education and experience are key for leaders who can sustain an organization in conflict.

<u>*New queens*</u>

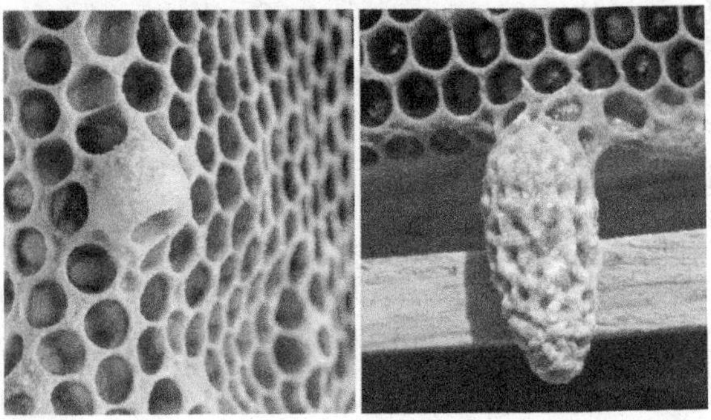

As mentioned earlier, when the hive is preparing to swarm, a new queen for the remaining hive is needed (see queen cups as part of the hive to the right). The development of the new queens is remarkably rapid, requiring only 16 days from the time the egg is laid to the moment when an adult queen emerges from her cell.[110] A beekeeper looks for the presence of queen cells during routine inspections. Besides the preparation to swarm, other reasons for queen cells could indicate a sick queen or possibly a dead queen. The stress of this event can result in catastrophe. A beekeeper could also create the need for queen cells. Beekeepers routinely attempt to 'split a hive,' to increase the managed hives. This effort has a 60-70% success rate – especially if the hive without the queen promptly nurtures a new queen in her place.

## The survival rate of relocation

Since not all honeybees are in the standard Langstroth hive (this is the box used in industrial and hobby settings mentioned in Chapter

3), survival after relocation is not that high. Think of the stress of finding a new home after moving 2-3 states from your last location.

Of honeybees in the wild, from Seeley's experience ~25% of the "founder" colonies (newly started by swarms) would be alive the following spring. In contrast, almost 80% of the "established" colonies (those already in residence for at least a year) would survive winter. Timing is also important for the success of the move – spring is better. "A swarm of bees in May is worth a load of hay, a swarm of bees in June is worth a silver spoon, a swarm in July isn't worth a fly."[111] Similar to changing residents—especially with 10-30,000 of your relatives, to a new home —now that's stress. Especially the debate before choosing which new home is the best.

*The debate over which home to choose*

The stress of a new home involves a debate with many hive participants, not just the queen, involved. A debate including hundreds of inputs (not 10-30,000, thank goodness!) from self-appointed investigative bees (previously foraging bees) who have taken it upon themselves to discover the perfect home (in their opinion) for the newly formed hive. Martin Lindauer, less than 100 years ago, discovered that when a honeybee swarm chooses its future home, it does so in a debate in which the nest-site scouts express their arguments in waggle dances (discussed in Chapter 6) positioned on the outside of the bees who have left the hive and are now in a temporary swarm ball waiting to choose the next home.[112] The precarious ball of bees, with the queen securely positioned in the center for warmth and protection, will remain in this holding pattern for 24-72 hours while selecting a future home.

We'll discuss this fascinating decision-making effort later in Chapter 5. But first, preparation for the move.

<u>Prepare to move</u>

First step – exercise the queen. She hasn't flown in months and is both overweight, laden with eggs, and out of flying practice. To help her lose weight, other bees will not let her stop walking. Worker bees will periodically push up against her and give her a quick jolt to keep her in motion. She needs to be in top shape for the stress of moving.

Besides prepping the queen, the part of the hive that leaves (~60% of the bee population) will gorge itself on honey. The bees are then homeless but not hungry, and their very demanding 24-72 hours move is going to require focus and sacrifice.

In fact, after the move to the swarm holding area, some forager bees (~100) voluntarily become real estate explorers or scouts. These scouts will work tirelessly to settle the debate on the new home as quickly as possible.

To summarize, the bees manage the stress of choosing a new home by:

- Exercising the queen for the flight
- Preparing future queen cells where new larvae can be fed "royal jelly" and developed
- Gorging on honey in preparation for the move
- Volunteering as scouts to find a new home

For bees, don't assume this swarming is a bad thing. A beehive can only reproduce in this way. One hive becomes two by swarming. The stress and possible failure of reproducing is great, but the honeybee population must face this stressful challenge to grow.

*What are the human parallels to relocation?*

Gayle Avery and Harald Bergsteiner, in their book *Sustainable Leadership: Honeybee and Locusts*, contemplate some interesting parallels to human leadership perils when confronting community stress. In step 9 of their leadership model, the beehive successfully demonstrates a healthy process for transitioning organizational leadership. They also look for industries that follow a similar model of careful organizational change and suggest similarities. As noted above, the steps of organizational succession in a beehive do appear very orderly except the final steps – a fight to the death by the remaining virgin queens to take control. Avery and Bergsteiner see a lot of positives in the beehive model and believe that companies that provide security of employment, who recruit for cultural fit, share information, empower employee participation, and allow for self-managing teamwork are sustainable.[113] The Avery and Bergsteiner book is a great place to see examples of industry using these concepts.

But how does one implement or utilize these lessons between bees and humans when addressing stress? Avery and Bergsteiner provide insight here as well. Since a swarming or relocation activity is a life and death struggle for the bees, the motivation, and the importance of working through difficulties – not just avoiding the stress – is critically important.[114]

*The stress of Killer Bees on swarming*

Unfortunately, not all common western honeybees swarm just once a year, with relatively large swarms occurring in the spring. In the mid-1950s, Brazil attempted to increase bee production by importing an African honeybee. The bees escaped containment and created what eventually became the infamous killer bee. These bees have wreaked havoc in the warmer climates ever since. Why? In the mid-'70s, Mark Winston, while in French Guiana, noted that the Africanized bees were swarming an astounding sixty-nine times per year, creating sixty-nine separate colonies.[115] Unhealthy? Yes. This attempt at increasing honey production resulted in a decrease in honey production and an angry, unstable hive unable to survive the winter. As mentioned earlier, swarming is good – but too much destroys the fabric of the honeybee community and results in a territorial bee only able to survive in warmer climates and unable to handle the stress of changing seasons.

Similarly, the occupations historically passed from generation to generation have faded in society. The Smiths no longer manage the iron works and the Bakers no longer oversee the grocery store. Our individual tendency to jump from occupation to occupation may provide some short term or intermediate gratification but beware of moving too quickly from either job or location. The idea of increased honey that resulted in the introduction of Africanized honeybees resulted in a relocation frenzy, destabilizing the healthy hives. Perhaps our own very mobile society could learn a lesson from such an un-sustainable practice.

<u>Summary</u>

In the play *Henry V*, Shakespeare has an apt quote on this marvelous insect: "…for so work the honeybees, creatures that by a rule in nature teach the act of order to a peopled kingdom." What do we learn about the way bees address stress? We learn the act of order when searching for the next home. We see the importance of timing for the future location of the hive. Finally, we learn from CCD about multiple impacts of environmental issues that, like a thousand cuts, seem to damage the very purpose of the beehive's existence, causing a mass suicide of the hive.

## Locusts and stress

Nothing creates more stress in the Saharan desert than the idea of a swarm of billions (maybe a trillion) locusts flying over the horizon and preparing to eat you out of your house and livelihood. But what about the locust? What stress can such an insect have? How do they manage stress?

### Pareto Optimality

For locusts, the economic term Pareto efficiency or Pareto optimality is useful to describe their existence. Pareto efficiency is a situation in which the reallocation of resources can make everyone at least as well off as they currently are. Stated in the negative, Pareto efficiency is where no individual or preference criterion can be made better off without making at least one individual or preference criterion worse off. The concept is named after Vilfredo Pareto (1848–1923), an Italian civil engineer and economist who used the idea in his studies of economic efficiency and income distribution.[116] Think of Pareto efficiency as the most effective operating condition any system could obtain. A locust swarm wreaks havoc with this "point" from a human perspective by their ability to upset consumption (a small locust swarm will consume the equivalent of enough food for 35,000 people every day). For the locusts, having the optimal number of consumers and the optimal amount of food to consume is the Pareto efficiency "point" pursued.

Of course, to locusts, the critical point does not include humans in its consumption distribution (and vice-versa). It joyfully consumes everything in sight with no concern for the future as it journeys toward this critical balancing point. The swarm grows and the journey to the next feeding ground seems like a stroll in the park. The stress occurs when this optimal point is exceeded and too many locusts are pursuing too few leafy vegetables. That stress leads to dramatic changes.

## Drought and Heat

Upon reaching the point of Pareto efficiency, other environmental stressors are now a concern but also an unusual blessing. The lack of water and vegetation following a surprising rain creates stress for humans but not locusts. In fact, this stressor favors locust growth.

The locusts flourish during drought, which provides several critical catalysts for population growth. The hot, dry weather weakens plant defenses and increases the nutritional value of the vegetation as sugars and other nutrients concentrate in the leaves. The dry condition also suppresses fungal diseases, which could reach epidemic proportions and devastate locust populations in wet years. Furthermore, the heat accelerates the locusts' maturation and development and gives them a boost in a difficult race against predators that inflict constant mortality on the dense bands of nymphs. So, the faster the young locusts make it to adulthood, the greater the proportion of the population survives to reproduce. Finally, in times of drought, lush vegetation gets restricted to swales and well-tended agricultural fields, so the locusts are forced to aggregate in these habitats, a behavior that initially generates and then sustains the coherence of both gregarious nymphal bands and adult swarms.[117]

As humans, we take for granted the availability of resources until a shortage occurs. In a recent presentation for the Alliance of Hazardous Material Professionals in Omaha, Nebraska, Dr. Custodio Muigana of the Center for Disease Control, noted that during his early years in Mozambique, the idea of waste was handled differently than in the West. Material of no use to one group was passed to the next. If no one wanted it, the children began to play with it. Once it ceased to be a child's plaything, it could be burned and used as fuel. Dr Muigana point was tongue-in-cheek, but shortages and difficulties don't always result in stress, many times they result in growth. The current emphasis on a circular economy is a healthy attempt to put this practice back into our daily practices.

A locust eyes

The head of a locust is broad and vizor-like with a cruel down-turned mouth and a pair of large protruding compound eyes capable of seeing from side to side through an angle of 220 degrees and up or down through 197 degrees. Such an immense range of vision makes the creature very difficult to catch. Such eyes enable it to look above, below, before, and behind with little to no head movement. These eyes are most evidently helpful in a swarm, when upwards of 120 million locusts per square mile are keeping stations with each other, and their protective purpose is probably secondary. The locust uses the joined antennae springing from the top of the head to make tactile inquiries toward any other locusts it may meet.[118]

When the locust is in the gregarious state, the increase in brain size is routinely attributed to its increase in sight and smell as well. Stressful situations result in strengthened, not diminished eyesight for the locust. In contrast, human eyes dim and narrow and become unnecessarily dilated under stress. In some cases, vision loss occurs. It's as if when we need our vision the most, that aspect of our five senses gives up first.

Under stressful situations human eyesight is perhaps the weakest resource we have. During an exercise with an attack helicopter battalion of the Texas Army National Guard, in 2002, I marveled at the protection placed around the pilots sleeping tent. To qualify to get into the cockpit and fly the Apache, the pilots must clock so many hours of sleep during any 24-hour period. Injuries from friendly fire can be traced to tired pilots with strained eyes. The Army knows the risk of eyesight under stress. Plan on restrictions concerning what you see and pause when you decide to trust your eyes in stressful situations.

## The locust's enormous fat tank

The locust eats the equivalent of their weight every 24 hours. Opening the abdomen of a locust reveals some evidence of this voracious appetite and the insect's need for it in the form of a soft yellowish amorphous mass of fat stored after digestion of the food materials. This is the insect's flight fuel and can keep it in the air for spells of up to seventeen hours – well beyond the capacity of any other insect.[119]

Consider the resources needed in a sustainable society in both the short and long term. When ~50% of the world lived on farms, farmers would routinely save a year of income not knowing whether the current harvest would be a boom or bust. Prepare resources for the unknown of future demands, like a hurricane preparation kit, a tornado shelter, or savings for a rainy day. The simple steps of preparing for short-term limited resources all help manage stress.

## Stress Increases Reproduction for the Locust

Among the more curious facts that have emerged from laboratory investigations into the whole field of sexual maturation of locusts is that if a locust has suffered from a traumatic experience of some kind of sexual maturity can occur earlier. If, say, it loses a leg at fledging, then it will be compensated by achieving reproductive capacity sooner; electric shocks and being tumbled around in a revolving test tube have also proved remarkably stimulating, but since neither of these is liable to occur on a widespread scale in a desert, other reasons have had to be considered.[120] Perhaps because one of the keys to locust swarms is the speed at which they reproduce, the traumatic experience is not a detractor but a benefit to their growth as a swarm. Apparently a catastrophic event doesn't slow them down, they adapt.

## Climate Change and Locust Growth

Climatic stress results in growth in the swarm. Roger Pasquier noted that sunspots and heat or rain spells following droughts, the consequence of any climatic upset affected locusts in a positive way. As the areas of desert vegetation expanded under the influence of rain, so would the desert locust's living space. As the same vegetation withered and shrank, so would the habitable patches, forcing the insects into each other's company. One condition would pave the way for their increase, and the other would control their density and likelihood of movement.[121]

Thus difficult weather conditions result in positive, not negative, impacts on their development as a swarm. Unlike honeybees, the locust's adaptability to weather is a road to success. Tension (which is what life is about for the locust, not homeostasis) results in a challenge, and a challenge results in growth. The stress of climate change prompts the locust to thrive.

*Rain and Stress*

Since desert locusts appear across such a large area where you don't expect them, no one has successfully eradicated the species. Consider the unique location chosen to lay eggs in a wadi with just enough water to provide for the eggs. Whether by divine design or not, these are precisely the conditions the locust requires for its own survival. It needs sand for its egg-laying. Like the plant seeds, it too, needs soil moisture equivalent to about an inch of rainfall for the eggs to germinate successfully. It too, must accommodate the most important part of its life cycle to the brief spell in which the wadi springs to life. But unlike plants, locusts can seldom wait to achieve full maturity but must eventually move on to seek new pastures. Just as nomadism is an essential part of man's defensive mechanism against the hostility of the desert, so it is with the desert locust.[122] Again, the stress of finding the right location, moisture, and soil type doesn't appear to limit the locust but provides the challenge of averting detection and growth.

Pesticides

Another stressor to the locust is pesticides. But even here, the locust has a unique defense. It's called chitin – a stress defender that fights pathogens. Chitin is the most important constituent of the cuticle or exoskeleton of the desert locust. The production of chitin is a continuous process and increases throughout the life of a desert locust, varying from about 1.7% of the fresh weight of a locust during the hopper stage to 2.2% in the young adult and 4% in a two-month-old adult.[123] Cockroaches also have chitin, which plays a key role in defense against pathogens. It's like armor. Remember the pet in the animated movie *Wall-e*? Why did the cockroach survive when everything else died? Chitin. Chitin protects the locust from numerous environmental stressors.

International Organizations and Predators

Living in the desert with an international organization in place to destroy your existence (the United Nations) as well as numerous other predators desiring to eat you, can also be enormously stressful. How does the locust manage this stress?

The straightforward answer is through its enormous adaptability to different environments. If rain and vegetation fail at the times required for breeding or for other reasons not yet fully understood, its numbers drop below an unpredictable optimum and the marauding locust retreats, yet it still survives. In fact, for the desert locust, there is no single place in the whole of the desert breeding areas of over five million square miles where it may be sure of finding its needed habitat. As a result, it can and will go anywhere (flying at 60 miles per hour for up to 17 hours in a single day). Locusts have extraordinary powers of flight to be able to traverse the vast distances required to reach where rain has fallen. At this point, they must then battle natural enemies—birds, lizards, fly larvae, and beetles, and almost all animals.[124] They fly downwind, they change their color to confuse predators, and they move in massive numbers. In short, to manage the external stress they have a choice to stay and die or to move along.

The changes in the insect's morphology (such as lengthened wings), behavior (such as the propensity to aggregate), and physiology (such as the delayed maturation of eggs) are all in preparation for migration. For locusts, crowding means an impending famine and migration into the unknown is preferable to sure starvation.[125] The stress of what comes next is secondary to death. Perhaps this variability to environmental stressors is the analogy. Difficulties can make us stronger.

*Serotonin and Stress*

In Chapter 2, I mentioned that Dr. Jekyll becomes Mr. Hyde based on the potion developed and taken. Some scientists have determined that serotonin (in humans, it's called the "feel good" chemical) may be the potion responsible for this behavioral transformation, being both necessary if behavioral gregarization is to occur and sufficient to induce it. In fact, data from Michael Anstey demonstrate a neurochemical mechanism linking interactions between locust individuals to large-scale changes in population structure and the onset of mass migration.[126] With serotonin in its system this creature adapts to stress as needed.

*Lack of food and Stress*

Food scarcity is the only identified stressor that slows the locust swarm. When the food runs out locusts can actually move back to the solitary phase. Through a few generations, the gold and black return to a camouflage of green or brown. The brain shrinks in size, and even the legs change in strength and length.[127] They minimize contact with others. They wait for the future opportunities to transform to Mr. Hyde once again.

But let's return to the locust. How does the locust, in the gregarious phase, transform to its earlier solitary state? As the population density declines due to a swarm's dispersal or high mortality, the cues that elicit the development of the gregarious phase diminish. The eggs laid by scattered adults tend to develop into unsociable nymphs. With low densities and solitary behavior, the nymphs rarely encounter one another (or other locusts' feces) which would prompt the transformation to begin again. And so, the locusts transform back into their solitary phase until such time as environmental conditions induce a surge of reproduction, a period of crowding, the rush of serotonin, the possibility of famine, and the compulsion to migrate.[128]

## Summary

Is stress bad for these insect communities?

Perhaps yes, but perhaps no. For the honeybee, environmental changes have resulted in a death from a thousand cuts (Colony Collapse Disorder). But for the Locust, environmental changes result in the very transformation that gives the insect, as a community, the fearful strength it must have to impact everything and everyone it meets. Similarly, the difficulties humans face appears daunting – and many times, they are. But the search for purpose and the "why should we sustain this at all?" seems to strengthen our resolve and our convictions. Frankl notes,

> Mental health is based on a certain degree of tension, the tension between what one has already achieved and what one still ought to accomplish, or his gap between what one is and what one should become. .... I consider it a dangerous misconception of mental hygiene to assume that what man needs in the first place is equilibrium or, as it is called in biology, 'homeostasis.' i.e., a tensionless state. What man actually needs is not a tensionless state but rather the striving and struggling for a worthwhile goal, a freely chosen task.[129]

With this as a lesson learned between the two insect communities, perhaps the stress confronted benefits both. Perhaps the right amount of stress is similarly beneficial for humanity as it allows us to sustain the growth and tension the stress provides.

Stress in the military

Preparing for stress is key to developing a well-trained military ready to win. The U.S. Army Ranger School creates a situation of growing stress over 62 days by removing food and sleep from the soldiers. The stress of operating with a lack of food and sleep creates not a weaker soldier but a stronger, more resolute soldier – one who sees the challenge as the normal tension to confront when pursuing a worthwhile goal. Learning to endure hardness is critical to success. Though Albert Einstein wasn't addressing the military, he also knew this concept and stated that in the middle of difficulty lies opportunity. Stress and hardiness are benefits to growth and success. Leonard Bernstein, the composer of *West Side Story* stated it best concerning stress when he said – "to achieve great things, two things are needed: a plan and not quite enough time."

So perhaps stress has its place, and homeostasis is more the enemy than the friend. Consider the arch. "If architects want to strengthen a decrepit arch, they increase the load which is laid upon it, for thereby the parts are joined more firmly together."[130] Throughout the history of the United States, the greatest leaps seem to be linked to the stress placed on our country. Perhaps, like the arch, stress isn't as bad as one may think and can provide the encounter – or the gateway necessary to confront future challenges successfully.

> *General Law – To fear change is to fear being challenged. To fear being challenged is to fear growth and new possibilities.*
>
> *Ty Howard*

 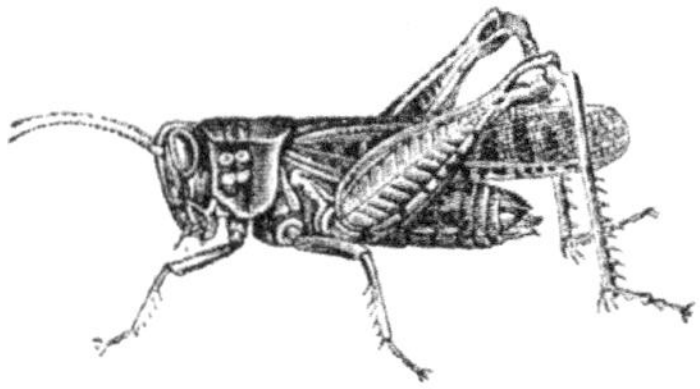

*Every accomplishment starts with the decision to try.*

– **John F Kennedy**

# Chapter 4: Work Allocation

In 1989, in violation of the Panama Canal Treaty, the Panamanian Defense Forces (PDF) blocked the single highway linking the Atlantic and Pacific Ocean. I had been stationed in the area for three years in the mid-80s and was familiar with defense of the area and the operations of the canal. I frequented the blocked road when traveling across the country. Now, as a Captain and a battery commander, my unit, C Battery, 2nd Battalion, 8th Field Artillery, was being deployed to the Atlantic side of the canal for additional defense.

The previous unit, B Battery, had developed two separate fire bases each consisting of three 105mm howitzers. Each position could cover both the Gatun Dam and the Gatun Locks outside of Colon. The positions were manned and prepared for use. My Battery took over these positions in July of 1989.

Interference on the road continued, and the deployed infantry brigade from the 7th Infantry Division prepared to demonstrate a greater willingness to use force. A proposal was discussed to airlift two artillery pieces to prominent hilltops along the route while an Engineer company moved obstacles off the road. The howitzers would be prepared for use, but the positioning of the weapons by Black Hawk helicopters would be more of a show of force than a useful system to engage the PDF. The landing zones, we were told, may be compromised, and small arms fire could be possible.

As I reviewed the plans and discussed the mission with the leaders of C Battery, I included a senior Staff Sergeant, experienced with artillery movement in Vietnam, in the discussion. This soldier had left the military for several years and returned in the late ‘80’s. As a result, SSG X was significantly older than the rest of the unit and I respected his opinion.

When I asked his advice on actions to secure the landing zone, he balked. His responses were hostile. I finished my review and reported back to brigade headquarters on what I thought would be the best approach.

Upon returning to the firebase, SSG X asked if we could talk in private. We did, and what he stated surprised me. His experience in Vietnam exposed him to units that didn't do well, and his reason, as he understood, had to deal with work allocation. As a Staff Sergeant his involvement in what was going to happen needed to come from the intelligent and thorough review of the situation available. His involvement was to execute. Asking for advice, he believed, was a sign of weakness that made him feel ill at ease. He did not want the question of tactical involvement presented to someone at his level. As he saw it, my job as the officer was evaluation and commitment – his job as a gun section chief was execution. Work allocation was important to him and I learned from his anticipation of roles, what work he believed had been allocated to him and how communicating a change in responsibilities, even just asking advice, disturbed him.

---------------------------------------------------

## Work Allocation and Bees

## Who does what in a beehive?

Most backyard bee hobbyists (like me) see the production of honey as the primary work of concern. The delectable 20-50 lbs. of honey I am able to collect from the hives is an example of the work accomplished by this active population of bees.

Yet the most important benefit from a human perspective is the pollination of the surrounding fields. Over 80% of all flowering crops require honeybees for pollination. However, the honeybee motivation is neither honey for our consumption nor pollination for our benefit but the development and storage of food for the hive to weather the change of the seasons and continue to thrive as a community. It's so easy to put man at the center of this effort – yet to study honeybees, we must evaluate why they do what they do and how they do it from their perspective.

Thus, the true work being performed by this fuzzy yellow and black insect is the maintenance and expansion of their home to sustain their society during the seasonal changes they will have to confront.

### *Busy as a bee*

When someone says you're busy as a bee, it's usually a compliment. It means you are focused, and your industrious efforts will bear fruit. Although bees are unlikely to have any cognitive understanding that they are focused, their behavior exhibits those aspects of concentration that result in highly efficient and effective work. Our human term would be "present in the moment," attentive to the one task being performed to the exclusion of any distractions.

But where did the moniker "Busy as a Bee" come from if it's not related to bees? Geoffrey Chaucer in *The Canterbury Tales* first used the phrase in the late 1300s, but it's misleading.[131] Worker honeybees are active occasionally, almost frenetically, but they routinely spend more of their lives resting rather than working. Bees spend up to two-thirds of their lives doing nothing.[132]

<u>*Resting as a bee*</u>

I can imagine that's a surprise. Worker bees only live for 4-6 weeks, and studies show that approximately 3-4 of those weeks are spent doing nothing. Of course, the queen bee must be busy. She is an egg-laying machine – depositing 1500-1800 eggs into the brood area every 24 hours. But the worker bee, not so much. Then how do they accomplish much?

<u>*Purposed as a bee*</u>

Bees make radical career changes every few days, going through seven to ten or more "professions" during their brief lives. Janitors, cooks, nurses, construction workers, guards, and farmworkers. Retraining lasts a few hours at most. Each job is accomplished with precision.[133] Multi-tasking just doesn't exist in the beehive. Perhaps as a lesson, it should be discouraged with humans as well.

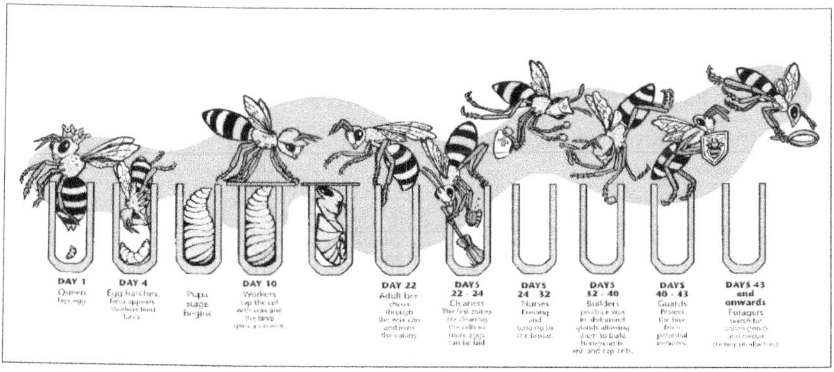


Bees move purposefully through tasks during their lifetimes. They begin by cleaning cells for a few days, then transition to feeding larvae from day five to twelve. Next, the bees receive nectar and pollen from the foragers and place it into various cells. After that, the bees mature to build a comb, fan the nest, and then guard the entrance. Bees finish their lives by foraging for five to ten days before wearing out and dying somewhere between twenty-five and thirty-five days of age.[134] One job moves seamlessly to the next with little or no training before they begin the work. Worker bees never procreate – they work and rest.

*Bees don't multi-task*

For humans, research conclusively demonstrates that each task suffers when we try to do more than one thing at a time; our brains become overwhelmed with information and choices. A classic example from psychologists Christopher Chabris and Daniel Simons addresses those who talk on a cell phone while driving. In their effort to multi-task, individuals on cell phones are four times more likely to have an accident than those not on cell phones. This is equivalent to driving while drunk—with 2,600 deaths and 330,000 injuries annually in the U.S. caused by distracted drivers using cell phones.[135] It's about focus. Bees may not be intelligent, but they are single-minded. There is a presence of mind that the bees have as they work, suggesting complete absorption in the one task they are performing at that point in their life.[136]

*Multi-tasking in combat*

Rushkoff mentions the danger of multi-tasking in combat. Drone pilots, those on the ground controlling the aircraft through cameras and computers, yet still participating in the war in Iraq and Afghanistan, have more PTSD issues than real pilots in the air. Apparently, the attempt to have humans act as computers and compartmentalize their wartime involvement separate from daily life doesn't work. They'll fight for a few hours in an office on a military base and then get into their car and drive home. Our brains can't take something as demanding as flying, fighting, and possibly killing others as just another day in the office. Humans can't compartmentalize or multi-task very well and we have serious problems doing it in extremely demanding situations like combat.[137] A sustainable community is wary of multi-tasking and its supposed benefits.

## Interaction of Wild Pollinators with Honeybee

So, if neither humans nor bees are successful at multi-tasking, are there different combinations of bees (different species) that can help get the work done? Do wild bees pollinate more efficiently than the standard *Apis millifera* that are trucked throughout the U.S. each spring and summer, or live in the beehive in your backyard?

### *The Almond Grove Experience*

The answer is yes and no. In studies comparing wild bees with honeybees, researchers in the almond tree fields of California discovered that honeybees tend to forage higher on the tree, and wild bees are evenly dispersed throughout tree canopies. Wild bees fly during stronger winds, colder temperatures, and more inclement weather than honeybees, even in light rain, making them a potentially significant contributor to pollination for almonds, which bloom early in the season, when high winds, cold temperatures, and rains are common. Perhaps most remarkably, the presence of wild bees enhances the effectiveness of managed honeybees. Since fruit set and yield are improved when almond flowers are pollinated by pollen from a different almond variety, almond trees are planted in rows that alternate between these varieties. This system benefits from the movement of bees between rows to affect the best pollination, but honeybees tend to move compulsively down single rows rather than cross the space between rows. Wild bees, on the other hand, move laterally between rows more readily than honeybees, but the presence of wild bees also induces honeybees to travel more between rows, improving their effectiveness as pollinators.[138] Just the presence of a competing example—wild bees with honeybees—results in a new practice for the honeybee population! But that still doesn't answer the question—which bees are the most effective pollinators?

*Wild Pollinators vs Honeybees – who's the best?*

During a 2013 study published by 49 authors in the journal *Science*, a definitive answer was reached. Based on 600 field studies on all continents addressing 41 different crops, the researchers found increased fruit set with wild-insect visits on all 41 crop systems (100%), but fruit set increased significantly with honeybee visits in only 14% of the systems surveyed, indicating that, when present, wild bees pollinate crops more effectively. The results "suggest that new practice for integrated management of both honeybees and diverse wild-insect assemblages will enhance global crop yields."[139] Work allocation, when wild insects are included, is more efficiently accomplished using this "non-managed" resource.

If true, why don't farmers worldwide move in this direction? For two reasons, one is ecological, and the second is financial. Meadowlands near farms, where wild pollinators would live, are either swallowed or integrated into the farm or closely managed as a park. For example, Britain has lost 97% of its meadowlands, more than seven million acres, since 1930. Most attempts to conserve wild bees also tend to be strongly organized by citizen-based groups, particularly in Europe and the United Kingdom, rather than government mandated efforts. In economic terms many farms are unaware of bees and view them as a managed tool brought onto and removed from their farms to supply a simple biological service. Thus, the short-term answer, when bee shortages occur, is to plant a variety that doesn't need pollination rather than to create a more bee-friendly habitat.[140] In short, we don't have an easy answer in the rural community to bring both the wild and managed honeybees to the table (so to speak!).

## Urbanization

Can we blame the problem of work allocation for pollinators on urbanization? Not really. A recent study discovered that bees in urban areas differ from bees outside urban areas. The world record for wild urban bees is held by Berlin, where 262 species were found, mostly in disturbed areas, rural patches, road edges, and backyard gardens. New York City—200 species. Sao Paolo, Brazil—133 species; Boulder, Colorado—88 species. Berkeley, California—74 species.[141] Bee species have been found to be more resilient in urban areas than most of us realize.

In fact, 45% of all vegetables in Hong Kong and 80% in Hanoi are grown on urban farms and gardens. Similarly, 80% of Singapore's poultry and 90% of the eggs and all the milk in Shanghai come from within those cities. In contrast, cities in the U.S. grow less than 10% of their own food. Urban food is also perceived as healthier as it's not genetically modified and is usually grown with fewer pesticides.[142] So what can we learn from this allocation of work? Perhaps honeybees' work can be augmented by something we don't even realize is available – wild pollinators.

Perhaps natural and unmanaged versus artificial and managed is not as stark a divide as we may think. Just as "city to wilderness" is a continuum rather than a border, perhaps we should also consider honeybees and the pollination completed as a continuum as well. Some say we should leave a minimal footprint on the earth, but perhaps it's more realistic to hope for a footprint light enough for other creatures to thrive while heavy enough to meet human needs for food, energy, and shelter. Balance is the desired endpoint, and that's not always easy to achieve. Perhaps the work allocation of pollination is more than just a honeybee-focused issue.

## Working ethically

The halo effect is when one trait of a person or thing is used to make an overall judgment of that person or thing. It supports rapid decisions, even if biased ones. How we feel and think about someone's character matters. Concerning honeybees, we tend to believe that the industrious honeybee must be an ethical and honest worker as well.[143] That is why state flags (Utah) and state emblems (15) abound with the honeybee. Unfortunately, such a bias is not accurate. Honeybees are also hoarders. Bees will take advantage as much as possible of a weaker hive and steal honey when the opportunity is right. The bees will mask the odor of the hive they are attacking to sneak past the guards and confuse the defense of weaker hives. Monitoring the entrance size of a developing hive is one of the first lessons a new beekeeper is taught. A weak hive needs a smaller entrance to protect. A stronger hive will have a ready guard force available.

The determination of who will be the queen of the colony is also something to consider. After a honeybee loses a queen, more than one queen will be developed as a replacement. The first queen-born attempts to kill those still in their cells by dashing over the combs in search of queen cells, chewing small holes in their sides, and stinging the occupant. If, however, two or more virgin queens emerge together, they will fight to the death, seizing each other and attempting to sting their opponent. The battling queen bees grapple and twist, each one struggling fiercely to implant her venom-laden stinger in her sister's abdomen. Ultimately, one queen succeeds, and the other, fatally stricken, collapses in paralysis; she falls from the comb and soon dies. The merciless sororicide continues until just one virgin queen remains alive.[144] The sustainable beehive needs a single queen, a central figure for the very important job of laying 1500-1800 eggs every day. Her ovaries contain the very life and future of the hive. However, how she arrives at this position is not one we should either emulate or use as an example of ethical behavior. Fortunately, she will live between 1 to 4 years performing the critical task of reproduction, providing stability for the hive.

## Worker bee strengths

Now the worker bees have a significantly shorter life but maintain unique biological advantages that, over time, play an important part in work allocation within the hive.

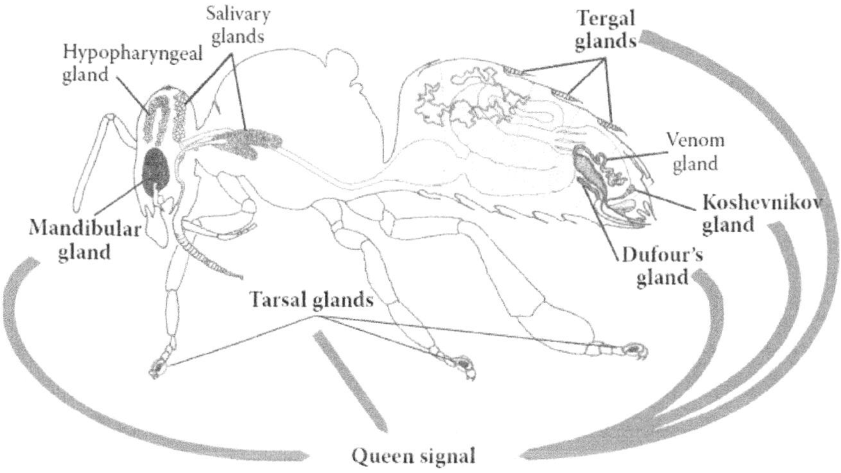


### *Wings and glands*

Because the young bees, after recently hatching from the pupa phase, have soft, untested wings, they can only provide the initial service of cleaning the hive. Similarly, their mandibular and hypopharyngeal gland are well suited to accept, process, and feed the young and older larvae.

*Wax glands*

As the hive bees mature to middle age, their wax glands develop, allowing them the crucial ability to build and repair the hive and cap cells which are in use for brood, pollen, or honey. During this crucial step, they complete the engineering task of construction as well as the production of the crucial construction material of wax. Do we both bring to the task the building material and the desire to build?

Railway Story

This is a true railway story.

Several years ago an extremely hot day, a crew of men were working on the road bed of the railroad when they were interrupted by a slow moving train. The train ground to a stop and a window in the last car-which incidentally was custom-made with air condition-was raised.

A booming, friendly voice called out, " Dave, is that you?" Dave Anderson, the crew chief called back, "Sure is, Jim, and it's really good to see you." With that pleasant exchange, Dave Anderson was invited to join Jim Murphy, the president of the railroad, for a visit. For over an hour the man exchanged pleasantries and then shook hands warmly as the train pulled out.

Dave Anderson's crew immediately surrounded him and a man expressed astonishment that he knew Jim Murphy, the president of the railroad as a personal friend. Dave then explained that 20 years earlier he and Jim Murphy had started to work for the railroad on the same day. One of the men, half jokingly and half seriously asked Dave why he was still working out in the hot sun and Jim Murphy had gotten to be president. Rather wistfully Dave explained, "twenty- three years ago I went to work for $1.75 an hour and Jim Murphy went for the railroad."[145]

*Foraging*

In the final stage of life, the brood-food glands and the wax glands are no longer useful, and the bee transitions into a field bee that will collect and forage pollen and nectar. Though this does require a second stomach, there is a need and an innate ability first to communicate information (through dancing) and second to process the honey in the stomach. Beyond this, these foraging bees seem very willing to sacrifice their lives to protect the hive! I have received several stings that can attest to this!

But how do bees know when to change jobs? Is it innate? Serial tasking based on age (and organ development) seems like a smart way to work, combining the benefits of specialization with those of being a generalist. But colonies must also have the flexibility to shift workloads if they are to respond successfully to changing internal needs and external challenges. How do bees, who are aware of only a small region in their colonies, know what to do without any colony-level command (i.e., no managers or supervisors)?[146] Many would assume the queen directs their effort. Queen pheromones are important and do communicate her health – but work instructions?

## Authoritative vs democratic approach

The queen communicates about her health through a glandular secretion called "queen substance," that workers contacting her pick-up with their antennae and distribute to all corners of the hive.[147] The only known dominion exercised by the queen is the suppression of rearing additional queens. This is the only direct decision she makes. She is not a dictator. Decisions in the hive have a decisively democratic flare about them. In fact, the effort necessary to swarm, which requires the queen to be in tip-top shape to fly to the new home, appears not to be entirely her decision either.

Initially, she indicates a future swarm (relocation) is needed by developing additional queens which she places in the queen cups already created by the hive. The queen undergoes physical changes in preparation for the move, but the physical exercise seems to be forced on her by other members of the hive.

*Getting the leader ready*

Her entourage of attendant bees feed her less and less and keep her walking. Her egg production declines, and her abdomen, no longer swollen with fully formed eggs, shrinks dramatically. Because of this the queen, who is normally too heavy to fly is fit and trim.[148] The entire process is one of teamwork. Are you preparing your leaders for success?

### Bee Conclusion

So, what can we learn from how honeybees allocate work? Work is allocated based on both age and physical aptitude over a short lifetime that is spent primarily communicating, not working. The bees work democratically with little to no oversight. The queen, though crucial for egg production, does not act as a dictatorial autocrat but as a fellow worker focusing on the overall success of the hive. Her preparation is even a hive responsibility.

## Work Allocation and Locust

Locusts are known for being exceedingly wise for remaining together. Proverbs states, "The locusts have no king, yet all of them go out in ranks." They march as if ordered to stay in formation as young nymphs and eventually flying as adults as if their route has been dictated from somewhere on high. From all external measures, their movements send a message of orderly, overwhelming yet destructive conduct. Yet what is their work?

## What is locust work?

To put it bluntly – their work is to consume and reproduce in a harsh environment. Each locust will eat its weight in vegetation every 24 hours. The gregarious swarm will increase in size approximately 20-fold every two months. The desert locust survives and thrives in an environment of extreme hostility continuing to plunder and grow fat.[149]

### *Reproduction*

Unlike the honeybee community, where reproduction is the privilege of the one queen, in the locust community, every female reproduces. A key difference between the solitary phase and the gregarious phase is this single function. One study tracked a solitary locust in Ethiopia, for which the incubation period lasted 14 days, the hopper period (from the hatching of the eggs to the beginning of adulthood) 38 days, the immature stage 45 days, and the breeding stage, ending in death, 30 days. 127 days in total, of which more than 2/3 were passed in a state of sexual immaturity.[150] For a human life comparison of 70 years, that would equate to sexual maturity of approximately 21 years. When the phase changes from solitary grasshopper to locust (i.e., the gregarious phase), this age of sexual maturity is increased by ~50%. Thus, a similar comparison to humans would add an additional 11 years to the reproductive possibilities of the insect – giving it several additional years (days) to reproduce.

*Food (protein)*

Both consumption and reproduction are unusual ways to qualify "work," but the final output of eating and reproduction is more locusts and less vegetation (i.e., successful products). As mentioned earlier, locusts consume the equivalent of their weight every 24 hours. Similar to grazing animals (cattle, sheep), the locust could and has in the past provided an ample supply of food to society. A single locust is approximately 50-60% protein which is more than beef. Few in Western society pursue locusts for this purpose, but other cultures and the American Indian have a history of depending on locusts as an important dietary staple.

## How to allocate work as a locust

One of the most remarkable things about a locust swarm is its cohesion. However strange the shapes it may take, however many the columns and curtains which may appear and disappear, it seems to be governed, above all, by the need to stick together.[151] Very little delegation or separation of tasks occurs. Locusts march as one—usually downhill—but as one. Locusts fly as a mass—from higher to lower pressure—but go basically in the same direction. Work accomplished occurs as a group.

*Color change – disturbs predators*

The work accomplished also occurs in a very bright and colorful uniform—yellow and black. The color seems to indicate a solidarity of effort or a unity in their new diversity. In the solitary phase, the locust is camouflaged to hide from predators, but in the gregarious phase, this changes. Why the change? Some speculate the colors provide the pause necessary to keep larger predators (i.e., birds, lizards) at bay. Others believe the locust eats poisonous plants and, with the change in color, confuses larger predators as to the health of this colorful meal. In either case, it seems to work. It wasn't until the 1920's that even humans associated this bright, colorful, gregarious insect with its passive, timid counterpart. Yes, the color change is helpful. Birds seem reluctant to attack the swarm. Instead, the gregarious locust, with their breadth of vision, can keep their eyes on one another. Is this important? Yes!

*Moving toward the center and waiting for the wind*

The locusts seem to have a strong affinity to always move toward the center of the group—to periodically move away but then return to the central mass and move, eat, and reproduce together. This occurs in the air as well. When photographed from below, double exposure (2 consecutive positions close together), observers noted instead of flying in the same direction, as one would expect, the swarm included large numbers of individuals and groups heading all over the place—continuously turning toward the center. Such a swarm is considered to be "waiting for the wind," which may itself then box the compass and bring the swarm back days later to its starting point.[152] The "waiting for the wind" concept is interesting. Locusts don't migrate. They appear nomadic, always moving toward what could be a more favorable area of consumption and reproduction. Downhill toward areas with more vegetation may exist. Downwind in search of low-pressure conditions where more rain may prompt growth.

If this "waiting for the wind" goes on too long, over too arid a stretch of country, the swarm may die, just as it must do if the wind consistently carries it forward, as often it does, into a country where its needs are absent. Yet so long as the swarm lives and keeps turning toward the center, so long will it appear to strive for the cohesion on which the best chances of group survival depend. This continual search for the center will be shown not only in the hoppers but among the flying swarms to which it eventually gives rise.[153]

The opportunity to take advantage of the natural tendencies of the world around us is a lesson in sustainability. The diet of the Inuit in northern Canada has an extensively high fat content whereas those of the islands in the Caribbean is significantly higher in protein. The eating habits of these people groups, based on the availability of sunlight and resources, prompt us to consider just what could we live with or without. We can allocate work based on needs but must also, like waiting for the wind, be cognizant of the resources and environment where we live.

## Reproductive success

An interesting study completed by the Tel Aviv University School of Zoology provides insight into the dynamic of gregarious male and solitary female production success.[154] Though not studied previously, this inter-phase interaction showed how the upsurge may find an amazing boost in size and strength when the two interact. The experiment showed that gregarious males were more aggressive and solitary females more accepting. Since solitary females produce larger pods (110 eggs vs 60-70), the move from an outbreak to an upsurge to an eventual swarm was accelerated. Although outwardly unchanged, the crowded solitary females now when these insect phases meet lay eggs that are biochemically predisposed to develop into nymphs with a predilection for aggregating rather than scattering. It's like jump charging the 'Dr. Jekyll to Mr. Hyde transformation. This tendency is reinforced as the growing nymphs are compelled to congregate in the shrinking patches of lush grasses caused by overpopulation, the onset of drought, or both. The constant jostling and odor of feces continue to induce endocrine changes that complete the locusts' transformation into the migratory phase.[155]

With the work of consumption and reproduction before them, males with an extended desire to reproduce and solitary females willing to work with the gregarious males and produce larger pods, an explosive second (upsurge) and dynamic third (swarm) generation is developed.

If the allocation of work (reproduction) finds interaction between the two locust phases and the larger pods prompt larger swarms, the locust has allocated work in a way the results in its exponential success – an even larger swarm.

## Summary

So, what do we learn from the way these two insect communities allocate work? For the bee, work is distributed, focused, and ends with a willingness to die, if necessary, for the hive. The output, honey, and the effort, pollination, are strongly encouraged by human interaction with this community insect. The structure (queen – drone – worker) appears dictatorial but is rather democratic and will be discussed in the next chapter on decision-making.

In contrast, the locust does not establish even the elementary steps of work allocation but appears to do all things together. Their cohesiveness and natural boldness (color change and movement) result in a fearsome, marauding band of insects that cannot be stopped until all vegetation is consumed.

Perhaps like the locust, humanity has at times created an unholy alliance between the relativism that results from having no cultural center and the habits of fashioning all worth based on consumption. One is a belief; the other is a habit. One comes from our internal world, the other from the external world. The consequence is a loss of centeredness. Like locusts, when we move toward a center that assumes no central principles, we are stripped of our ability to discriminate. We even lose our ability to distinguish good from bad taste, good from bad behavior, worthy ideas from those unworthy, and things that are serious from those merely trivial. We move toward the center not knowing where the center is.

A cultural center helps humanity make sense of work allocation. Do we employ children or the elderly? Who oversees the education of children, the parents or the government? Do we allocate work or allow individuals to choose? Without central principles, relativism as a "belief" reigns supreme.

Consumption is the habit we've learned as we buy, consume, and produce so much to stay up with the culture we occupy. Fast food has resulted in a relationship change in having meals like never before. The family table changes from the allocation of work and the sharing of values to something almost never visited. A sustainable society needs to consider the consequences of this unholy alliance and the impact it will have.

The family has been called the "giant shock absorber" of society – the place to which the bruised and battered individual returns after doing battle with the world after a day of work. The one stable point in an increasingly flux-filled environment. Pessimists tell us the family is racing toward oblivion – but seldom tell us what is taking its place. Consider the family the center.

And what about the reproductive success of the locust? Can humanity learn anything from this example? Perhaps the gift of children and seeing each one as just that, a gift could be the lesson. Do we believe this? Is the continuation of our species a benefit or a detriment? From a *mimetic* view that question would never have been asked. From a *poietic* view it's being asked all the time.

*General Law – There is no power for change greater than a community discovering what it cares about and how it allocates work.*

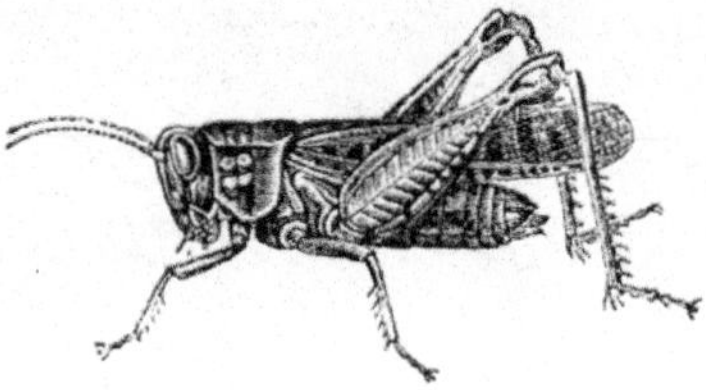

*Nothing is more difficult, and therefore more precious, than to be able to decide.*

– **Napoleon Bonaparte**

# Chapter 5: Decision Making

In 1997, my family was securely nestled into a small farm near Goshen, Indiana. The house had previously been owned by an old-order Mennonite family with an expansive kitchen and a wonderful red barn. With numerous animals to care for, my wife and children were very busy during the day while I worked in a nearby pharmaceutical plant.

Unfortunately, the production lines used at this plant were scheduled to be transferred to oversea operations, decreasing the workforce. The once rather sizeable location lost more than 50% of its employees in a two-year period because of attrition and lay-offs. As an engineer, I could see corporate support for the operation waning as future projects were suspended. As a result, I began looking for a transfer.

In Baytown, Texas, I found a rapidly expanding chemical plant searching for someone with my skills. The move was a promotion, and the new work which appealed to me appeared very challenging. Yet, my family felt closely connected to the community and farm in Goshen. What decision should I make?

I visited the Texas location and looked at the available housing. The team I would be working with were very friendly, and I could see myself easily fitting in. I decided to move.

Yet when I returned to Goshen and announced my decision, not only was my family upset, but many of my friends were as well. Why didn't I ask their opinion about the move? Why didn't I consider the possible jobs in the Michiana area? Why didn't I wait for the input of others before making such a big decision? The concern of "I, me, my" forgot about the "we, us, them." I damaged a lot of relationships in that move and learned one of the most important lessons a good friend told me before leaving that part of the country—there is wisdom in a multitude of counselors—use them all before making a decision.

---

What is decision-making? Simply stated decision-making is a twofold process: first, identify the possible alternatives, then choose among them.

Which decision-making efforts have an impact on sustainability? All of them. There are no small decisions. Benjamin Franklin didn't create the importance of this concept, but he did publish it in his 1758 edition of *Poor Richard's Almanac* when the American colonies were not seeing eye-to-eye with the English Parliament.

"For want of a nail, the shoe was lost.
For want of a shoe, the horse was lost.
For want of a horse, the rider was lost.
For want of a rider, the battle was lost.
For want of a battle, the kingdom was lost, And all for the want of a horseshoe nail."

But group decisions and societal decisions are unique.

Unlike Rodin's Thinker, the notion of a lone individual churning out ideas and making critical decisions in isolated contemplation is not the norm for societies. Kevin Dunbar recorded hundreds of hours of video of scientists working in labs. When he analyzed the footage, he saw that the vast majority of breakthroughs did not occur when the scientists were alone staring into their microscopes or pouring over data but rather when they were engaged with one another at weekly lab meetings or over lunch. Thinking is not just a personal activity but a collective one.[156] Group decision-making seems to benefit all and frequently presents the most equitable of decisions.

Concerning sustainability, the critical decision being made by honeybees and locusts addresses the survival of the next generation. How are those decisions made? Who makes them? When made, how successful are they?

## The bee brain

The size of a normal honeybee swarm is 3.5 pounds of bees, just like the 3.5 pounds of neurons that make up the average weight of a human brain. Like the human brain—with other human brains in near proximity—a group of bees routinely make first-rate collective decisions.[157]

But shouldn't all honeybees in the same hive who are genetically the samethink alike? They have the same mother and 99% of them are female. They live in very close proximity to one another. All these similarities should negate any variation in approaches. Yet even though the bees come from the same mother, the queen has inputs from multiple fathers, resulting in genetic variation. Not every bee is the same; some are more likely to collect pollen than others based on having lower thresholds for the factors that stimulate pollen collection. Some may be more apt at collecting water, nectar or creating propolis. In fact, genetic diversity turns out to be a key element for effective colony decisions as it buffers the colony from reacting too weakly or strongly to a single stimulus.[158] But the queen makes all the decisions, right?

## The decisions that are crucial

The queen does make decisions on whether to sire additional queens, but other critical decisions for sustainability are made by a much larger group – the entire hive. The honeybee dance used by foragers to indicate the location and quality of pollen and nectar sources is famous. The forager's dance communicates the location of pollen and nectar based on the intensity of the information exchange. The foragers' dance motivates other bees in the hive in the direction of where to find the best resources for feeding the bee community. From this exchange, some forager bees decide to check it out. Let's briefly look at the waggle dance.

### *Food source*

When a worker discovers a good food source, she will return to the hive to perform a waggle dance to let her nestmates know where it lies. A bee performs the waggle dance to share important information.

The direction of the food source is indicated by the direction the dancer faces during the straight portion of the dance when the bee is wiggling. If she wiggles while facing straight upward, then the food source may be found in the direction of the sun. As she changes the angle of the central portion of the dance, she communicates the direction left or right of the angle from the sun.

*Round dance*

The round dance is used for food sources 75-500 feet from the hive or closer. After distributing some of her new-found nectar to waiting bees, the scout will begin running in a small circle,

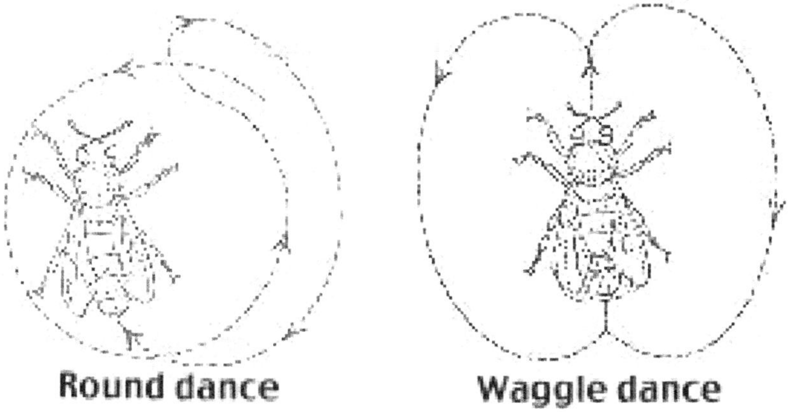


switching direction every so often. After the dance ends, food is again distributed at this or some other place on the comb, and the dance may be repeated three or more times.

The round dance does not give directional information. Bees who are encouraged to forage after a round dance, fly out of the hive in all directions, searching for the food source they know must be there. Odor helps recruited bees find new flowers in two ways; first, bees watching the dance detect the fragrance of the flower left on the dancing bee. Next, the scout bee leaves odor from its scent gland on the flower that helps guide the recruits to the source.[159]

*The Waggle and sickle dance*

As the food source becomes more distant, the round dance is replaced by the waggle dance. There is a gradual transition between the round and waggle dance, taking place through either a figure-eight or sickle-shaped pattern. We'll discuss this in greater detail in chapter 7 on communication.

## The decision to divide the colony

The selection of a new home is, perhaps, the most crucial decision honeybees make, and the waggle dance is also used in this effort. Because a nest site that's too large or small, drafty, or wet could be fatal, this becomes a key sustainability decision for long-term survival. The site must be large enough to build sufficient comb to rear brood and store forty to sixty pounds of honey over the winter. It must not be so large as to require excessive energy to heat in the winter and cool in the summer. A forty-liter volume is ideal. Also, a fairly small entrance that can be defended from predators would keep the colony safe. Finally, an entrance at the bottom of the nest facing south would aid with thermo-regulation.[160]

### *Locating the new home*

When the decision to move occurs, an amazing group of volunteers (previous foragers) take it upon themselves to become real-estate hounds or scouts. These scouts begin searching for the future home of approximately 30-50% of the existing hive population. They explore the countryside in search of the right size, direction, height, position of entrance, volume of the cavity, and for an additional benefit – the existence of previous combs.[161]

As Seeley notes in *Bee Democracy*, scouts discover about twenty-five potential nest sites on average and bring that information back to the swarm (a group of bees hanging from a branch or holding area some distance from the original home). In this location, they debate the sites' qualities, come to an agreement on the best option, and then lift off as one mass to fly and inhabit the new home.

*How do the bees make their new home decision?*

First, the bees gather information in the accumulation phase. Second, debates end with all or nearly all the dancing bees advocating just one site, that is, reaching a consensus. Third, the bees support the decision-making process and, with the scout bees initiating the move, fly off in unison to the new home. The process involves dozens or hundreds of individual bees. When all is complete surprisingly, only 3-5% of the bees in a swarm participate in the dance debate, yet all, once a decision is reached, risk their future as a swarm.[162]

Seeley outlines six steps in the decision process:

1. The sensory apparatus of a swarm is a sizable population of scout bees.
2. The scouts collect sensory information (details about the potential nest) for several hours or several days.
3. Each scout makes an independent evaluation of the site they explored.
4. Scouts report on a site (waggle dance—the greater enthusiasm and confidence in the finding is indicated by the repetition of the announcement on the body of the swarm) and recruit additional scouts to the site.

5. Scouts reduce their dance responses over time (more about this below).
6. Scouts may adaptively choose between exploring versus exploiting in the debate.[163]

The debate is like the dance mentioned above concerning the source of pollen and nectar. Potential nests that seem superior receive more vigorous dances than those that are mediocre or of average quality. Both the rate of the dance and its duration are greater for better nest sites, thereby communicating the scouts' evaluation of quality. For inferior sites, these scouts perform lazier dances and recruit fewer new scouts, gradually reducing their own visits. With time they stop dancing for that nest altogether.[164]

What information does the scout collect, and how does it collect that information to make a recommendation? As mentioned earlier, size is a critical concern. If the hive is too large, it would be difficult to heat, cool and defend. If it's too small, it won't hold enough honey to get through the colder seasons. To determine the size, the scout performs a fascinating walk evaluation. In fact, Seeley mentions that a three-dimensional reconstruction of the walking paths of individual scouts reveals, when the inspection is finished, that scout has walked 200 feet or more around the inside of each cavity and has covered all its inner surfaces in order to assess the important aspect of size.[165] Of course, the walk helps to also assess dampness, opening, height above ground, etc.. In the end, the scout has a considerable amount of information to help add to the debate.

## Crowdsourcing

Is what the bees are doing "crowdsourcing"? Crowdsourcing was defined by Francis Galton, who used the example of guessing the weight of an ox at a local fair in 1906 as his most famous example. He noted that a total of 787 guesses were collected at the fair. The result? The average of all guesses was within a few pounds of the correct answer.[166] Crowdsourcing is the practice of obtaining information or input into a task or project by enlisting the services of many people. Bees appear to demonstrate something similar. Potential nest selection follows similar rules of crowdsourcing in that:

1. Each scout bee (~100) decides on what location to inspect independent of all the other bees.
2. Genetic diversity and variance in a colony's scout bees do exist providing unique inputs.
3. The decision about nest selection is decentralized and based on information from each bee's exploration.
4. The individual bee actions aggregate into effective colony information interaction and nest selection.[167]

*Is crowdsourcing good?*

James Surowiecki, a business columnist for *The New Yorker* and author of *The Wisdom of Crowds*, illuminates why crowds are more knowledgeable than individuals. He describes how the collective intelligence of groups leads to the best decisions and generally provides performance superior to that of a single or a few decision-makers:

> We're programmed to be collectively smart. Groups are remarkably intelligent and are often smarter than the smartest people in them. The idea of the wisdom of a crowd is not that a group will always give you the right answer but that, on average, it will consistently come up with a better answer than any individual can provide.[168]

The list of crowdsourcing projects is very extensive (see the list of crowdsourcing projects on *Wikipedia* – there are over 120+ with notable successes). Some would use this argument to promote full democracy – in which every citizen votes on every issue – in any country.

*Is crowdsourcing dangerous?*

Yet not all are as positive as Surowiecki. Henry David Thoreau stated that "The mass never comes up to the standard of its best member, but on the contrary degrades itself to a level with the lowest." Friedrich Nietzsche was even more negative about group intelligence when he wrote, "Madness is rare in individuals—but in groups, it is the rule." Mob rule (the worst case of crowdsourcing) is never given a positive light in society, and those that have witnessed the frightening response of a crowd following a losing soccer or football contest can attest to this fact. Nevertheless, the reality of honeybee swarms making good decisions shows us that there really are ways to endow a group with a high collective IQ.[169]

When seeking sustainability lessons concerning decision making one needs to see both the opportunity as well as the drawbacks of crowdsourcing. The collective IQ is considerably higher, but humans tend to irrationally hold a position based on emotional connection when in groups. The larger the group the more infectious the irrational hold on a position occurs. Even telling lies seems to increase (see studies by the National Institute of Health on the impact stress and groups have on the *Amygdala*). Madness, as Nietzsche says, is the rule within mobs.

Quorum vs. Consensus

So how do you get 30,000 bees to decide once the debate is finished? Since consensus building is sometimes the basis of democratic decision-making, such as trial juries, Quaker meetings, and groups of friends, most would assume this to be the chosen approach. Unfortunately, what is common when a human group ends a debate, election, or other democratic process routinely results in factions strongly divided in their preferences. At this point, the group must invoke some formal decision rule, such as, the majority rule or a weighted-voting system, to translate its split vote into a single choice. This kind of group decision-making has been called "adversarial democracy" because it arises from individuals with conflicting interests and preferences. In contrast, the group decision-making of a bee swarm is "unitary democracy" since it involves individuals who have congruent interests (i.e., choose the best homesite).[170]

To decide, the hive finds a balance using a quorum. Recognizing a valid consensus with bees is very difficult, but not with a quorum. A quorum is when enough of the parties that have an interest are present to create a decision binding even over the wishes of those not present. The 100+ scouts, dancing and debating on the surface of the swarm hanging from a branch in the woods, are that quorum.

Thus, the swarm doesn't use consensus sensing but quorum sensing. Why? For two reasons. It enables a swarm to strike a good balance between speed and accuracy.[171]

*Speed vs. accuracy*

Speed is of the essence in the decision. All swarm participants are using limited energy—like fasting—during the move. The worker bees eat as much honey as they can prior to the move, knowing that their next meal may be a few days away. Also, the queen and her opportunity to lay eggs are now on hold, and only a quick relocation to a new nest can get everything started again.

Accuracy is also crucial. If a decision is made too quickly, an unsafe and possibly fatal location may be chosen. If the swarm waits too long, energy runs low for the workers and weather conditions (rain, cool nights) will kill an excessive number of bees. Fortunately, when making the choice the leakage of information helps reduce the options. Leakage is the key ingredient allowing for a balance between speed and accuracy for the honeybee decision.

*What is leakage?*

Simply stated, leakage is the drop in excitement, intensity, or willingness to repeat a recommendation of what the scout bee found or believes to be the answer to the question—where we should go to nest. Information leakage is a critical part of the debate. Remember that the five key attributes of a perfect honeybee nest are: cavity volume (40-liters is best); entrance height (15 meters above grade is best); dryness; entrance size (relatively small); and direction of opening (south). In Seeley's experiments, bees that discovered 40-liter boxes resulted in strong dances (35 dances lasting 85 seconds) and the bees that identified 15-liter boxes performed weaker dances that, on average, contained only 14 circuits and lasted only about 45 seconds.[172]

As the entomologists watched, they discovered that each subsequent waggle dance for the returning scout was reduced in strength by 15. In other words, after returning to the hive after a follow-up inspection of the possible nest, her dancing over the swarm was reduced by 15 circuits.[173] It appeared that she was losing interest in the debate or at least in the willingness to win others to her position that what she found was best.

If you consider the two nests mentioned above, proponents for the 15-liter box will end their involvement in the debate after one trip to the nest. However, the advocates for the 40-liter option would continue their promotion at least three times as long. Thus, the hive would see, by the mere fact of the reduction in the intensity of the debate, the favored location.

*What is the leakage change rate?*

Seeley looked at what would happen if this leakage rate were raised or lowered from the average 15 dance circuits per trip back to the swarm. Leaked more slowly, the bees came up with a rapid but less accurate decision. Leaked faster, the models showed that the swarm made a less rapid but more accurate decision. Thus, bees have a good balance between speed and accuracy in choosing their homes with a 15-dance circuit leakage following each trip.[174]

Truly optimal decision-making for humans is rare. Decision-makers pay costs in time, energy, and other resources to acquire and process information. These costs usually preclude making the decision using all the relevant information.[175] But the uniform leakage of intensity of support helps honeybees find the proper balance between speed and accuracy and not to become too enamored with a single option. What can we learn? Honeybees appear to be willing to surrender an option much more willingly than mankind. Perhaps this is a strength necessary for sustainability.

*Leakage in human decision making*

When do humans give up on a recommendation in a debate? Humans will usually (and sensibly) give up on a position only after learning of a better one. In contrast, the bees will stop supporting a position automatically (after it retires from the debate as a result of leakage). Max Planck stated, "A new scientific truth does not triumph by convincing its opponents and making them see the light, but rather because its opponents eventually die, and a new generation grows up that is less familiar with it."[176] Such an approach is what we follow but perhaps that's not the best way to retire a decision—wait for the death of the opponents!

Of course, for human society, we do rely on groups to be more reliable than individuals when it comes to making weighty decisions. This is why we have juries, boards of trustees, blue-ribbon panels, and nine justices on the U.S. Supreme Court.[177] Yet all these groups are small in number. Is there a size beyond which humans should be wary of the honeybee decision-making lessons? Remember mob rule.

The Rule of 150

Malcolm Gladwell, the author of *The Tipping Point*, would say yes. "The Rule of 150 says that congregants of a rapidly expanding church, or the members of a social club, or anyone in a group activity banking on the epidemic spread of shared ideals needs to be particularly cognizant of the perils of bigness. Crossing the 150 line is a small change that can make a big difference."[178]

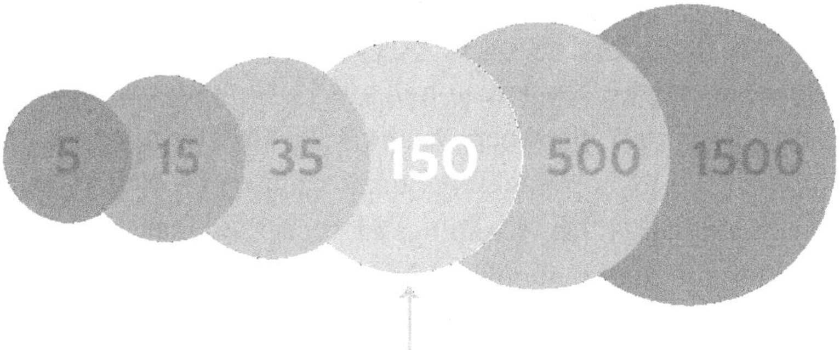


**Dunbar's Number**

***the max number of relationships a person can maintain***

Gladwell points out that as an organization's size increases, workers' ability to make independent decisions becomes more difficult but more important.[179] At a point above 150, we apparently lose the breadth of social contact to make optimal decisions. Robin Dunbar, the individual attributed with this idea, stated that the limit is derived from our brain's ability to maintain memories of 150 people and the time necessary to devote to the group in order to to keep 150 relationships viable. Dunbar stated that 150 people is the point beyond which members of any social group lose their ability to function effectively in social relationships. Roman military companies (centurions) to current military units (companies, batteries, troops) are typically 120-180 individuals. Neolithic farming communities are known to be around 150 people before they split into a separate community. The average parish size for the Hutterites and Amish denominations known for espousing communal living is about this size.[180]

Gladwell uses Gore and Gore Industries, the manufacturers of Gore-Tex, to illustrate. Gore displays an extreme form of self-management known as organic leadership. It is difficult to implement but very powerful when done well.[181] Simply stated, through trial and error, they have devised a model of office design whereby if the office grows to exceed 150 employees, they will start another office elsewhere until it hits 150 employees, and if that hits 150, they start elsewhere again. Like the beehive, a certain dynamic of size and optimal functionality appears with humans. As we frequently tout the technological successes that allow us to easily cross international boundaries and have work teams in every area of the country, perhaps we should still learn something from the rule of 150. Our ability to make important decisions as a group has limitations based on size.

## Do honeybees always make a perfect decision when looking for a new nest?

No. Seeley notes that honeybees sometimes don't agree, and the hive swarms divide in flight since the results of the debate weren't clear. How about the first choice/recommendation of the new home – do honeybees get this correct if they stay with the first option? No. Seeley noted that "from eavesdropping on the scout bees' debates, the early dancers tend to advertise losing sites."[182] Thus, we can learn sustainability from honeybees and their balanced approach to nest selection—but the approach is real and can fail.

Of course, the decision-making style of the honeybee does collect significantly more input than the locust, which we will discuss next. For the locust, the deliberation of inputs doesn't appear to occur at all, but they still are tremendously successful as they move across the landscape and devastate an area. Just as Shaw admonishes us to "Go to the honeybee, observe her ways and be wise", let us now turn and learn something from the locust and be wise.

## Decision-making and locust

The desert locust is the winged big brother of its fellow acridid, the familiar grasshopper of North American meadows, and quite often leads much the same sort of life as those grasshoppers. It is normally quiet and reclusive. It is a lonely insect attempting to hide in the meadow and stay away from everyone and everything—including its own kind.

In the solitary phase, avoidance seems to be central to all decision-making. The desert locust will elude everything. As a result, sexual interaction is delayed stopping the generation of millions to billions of locusts. In the solitary phase, the desert locust is the ultimate social distancer. Not only will it give other locusts six feet of clearance, it will do its very best to hide completely if another of its kind is in the vicinity. For protection, the solitary locust is green or brown in color. It takes flight at night and does its best to remain inconspicuous.

In the gregarious state, all that changes. The locust makes decisions based on togetherness with a flash of daring in almost all it does. Why do they stay together? Not for hunger. Not for stress. In some mysterious locust fashion, they cheer themselves on to stay together as they march and as they fly. In this gregarious group, the desert locust must move to eat, but they don't devour an entire plant. They merely take a few bites and move on. The desert locust rarely consumes everything.[183]

The locust appears to be audacious and adventurous. They decide (by nature) to obtain a bright yellow and black color which easily surrenders their location to a predator. Instead of flying solitary and at night, they decide to fly during the day, once again, surrendering themselves to predatory birds. Moreover, they purposely eat poisonous plants, which, when digested by the many predators attempting to eat them, results in the poison passing to the predator. Like colorful kamikaze pilots, the gregarious desert locust allows some in their number to become sacrificed so that others can succeed, grow, and reproduce.

## The decision – to eat and reproduce

Reproduction for the gregarious locust is the decision of greatest concern. It appears to be a critical driver for all that they do. In the gregarious stage, they appear solely to live to eat and to mate. How long do they copulate? Three to fourteen hours at a time, with multiple interactions throughout the day and evening.[184]

Who instigates this driver in their existence? Apparently, the male matures first. His yellow color change shows in the tip of his abdomen, which also appears more pronounced or brilliant among all the locust when he is ready to copulate. Second, the female locusts' maturity is hastened by a skin odor given off by the male. The odor affects other males as well, so the process of maturation of the entire swarm, numbering many millions, is speeded up and usually completed within a week. Finally, it is believed that vision, the most powerful factor in holding the swarm together, also plays a key part in this sudden burst of sexual unanimity.[185]

Are locust eyes key to mating?

In an article in *Environmental Entomology,* the mating behavior of the desert locust was studied in a laboratory, exploring the hypothesis that long-range communication signals may be operative for mating in the solitary phase. As mentioned in the previous chapter, other groups had discovered that gregarious males and solitary females appeared to copulate most frequently and successfully. What did these female locusts in the solitary state have that their gregarious counterparts lacked? Is there some long-range sex pheromone in this phase or is it sight?

In the experiment, before mating, males actively scanned the visual field, frequently cleaned their antennae and compound eyes, and jumped toward the females. The females, in response, moved slowly in front of the males and vibrated their entire bodies. Males assaulted and grasped the females, which at first tried to kick off the males using their hind legs but eventually submitted. In wind tunnel experiments with hidden females, males spent ≍33% of their time searching, and >80% were able to locate females, whereas, in the absence of the hidden female's stimulus (i.e., pheromones), they spent little time (<5%) in searching and remained in a resting state near the release point. In the presence of hidden females, the males also scanned and jumped more frequently, whereas they did not actively scan or jump in the absence of hidden females. When males were held in wire screen cages, and females were released downwind, the females did not respond to the males, indicating that males search for females when mating and not vice versa. Blinded males showed no interest in feeding or mate finding. These studies affirm, for the first time, the emission of a sex pheromone by the solitarius females attracts males with sight being a key input.[186]

Perhaps a bit complex but note the impact of blinding the males in this experiment. Once blinded, they showed no interest in feeding or mating. Pheromones did prompt a reaction between the male and female, but without sight, the locust loses the will to live. Its eyes were key when making decisions.

Our human eyes are also crucial when making decisions. One of the repercussions of the current overuse of social media is a lack of experience of what to do with our eyes in face-to-face interaction. The amount of communication that is nonverbal may be up for debate (somewhere between 30-80%, based on the source). However, without well-trained eyes, communication becomes very difficult and decision making is retarded.

## The Perfect Environment

The ground conditions for reproduction are important as well. We think grasshoppers and locusts are creatures that live above the ground, but they spend more than three-quarters of their lives buried in the earth. It is no wonder that the chemistry, moisture, texture, and depth of soils are so critical to these insects.[187] The decision of where to position the eggs is perhaps the key to the locust's success and also the bane to humanity's problems with locusts. What's the best place to plant locust jr. for his development? A breeding place not too far, in locust terms, from the desert fringes, adequate moisture ensuring plentiful vegetation, and far enough off the beaten track that observation is unlikely.[188]

When the 'right spot' has been identified, numerous other female locusts place their eggs together, resulting in a future explosion of young locusts in their first instar. All the pods hatch within hours or days of one another. Some hypothesize that a mechanical stimulus is at work. The ground actually erupts with these insects.[189] The United Nations Locust Watch states:

> The desert locust females lay eggs in an egg pod primarily in sandy soils at a depth of 4-6 inches below the surface. A

solitary female lays about 95-158 eggs, whereas a gregarious female usually lays less than 80 eggs in an egg pod. Females can lay at least three times in their lifetime usually at intervals of about 6-11 days. Up to 1,000 egg pods have been found in one square meter.[190]

If you do the math, that would result in 95-158,000 locust eggs in one square meter! With this accomplished, a 20-fold increase in the locust population will occur every 3 months.

## The Decision of location – Nomadic vs Migration

With the major motivation for decision making, reproduction, addressed, what is the decision process for relocation? R. Pasquier, answered this question in 1950. The locust doesn't migrate – it's nomadic. Migration is beyond the capabilities of a locust.[191] They displace, moving with the wind or with the contour of the land. From high pressure to low pressure. From high ground to low ground.

Thus, we have nomads. A very large nomadic society – but a nomad. The decision for where to move next appears always to be the physical pathway that provides the least resistance to forward motion—a choice to always take what appears, in the short term, the easiest route.

From a sustainability perspective, the path of least resistance is successful for the locust. They pursue food and pleasure. Their path leads them to more vegetation, and they continue until all possible resources are consumed and their next generation begins to die. In the short term it appears to be a sound strategy but if left unchecked results in death to the society.

## Uvarov and complexity theory—Decision by the numbers

Do numbers make a difference in decisions? Boris Uvarov, the Russian entomologist who discovered the phase change in locust, was a scientist far ahead of his time with regard to conceptualizing the natural world. The modern science of complexity with all of its attendant forms (chaos, catastrophe theory, self-organized criticality, etc.) constitutes a beautiful and compelling mathematical formalization of the concept that Uvarov articulated for biology: "Quantitative differences beyond a certain point pass into qualitative changes." Or, as so concisely expressed by the physicist P.W. Anderson, "More is different."[192] Pursuing the path of least resistance, as these locusts do, shifts everything. Their quantitative difference results in qualitative changes everywhere. Their supposed lack of a decision as to where to go is a decision, and the sheer size of a locust swarm makes a difference – a big difference. Concerning sustainability, pursuit of the path of least resistance will result in something qualitatively different.

## Summary

Bees – lessons to apply

Seeley includes his book on *Bee Democracy* with a five-step process he believes humans should use to improve their own decision-making. First, compose the decision-making group of individuals with shared interests and mutual respect. The foragers, or scout bees, with their seniority and abilities, have the confidence of the hive and the experience others lack. Logically, they are the best choice of bee-type to make up the scout group. Second, minimize the leader's influence on the group's thinking. Even though the queen is present in the hive and will, through her pheromones, direct the heart of the moving swarm, she neither participates in nor decides the location of a future nest.

Third, seek diverse solutions to the problem. As many as 100 volunteer scouts will step to the plate and look for a new home. The debate is successful as long as diverse solutions have been given the airtime needed. Fourth, aggregate the group's knowledge through the debate. The intensity of exchange on the swarm ball hanging from a tree is a debate, and the bee community must sit and listen.

Finally, use quorum responses for cohesion, accuracy, and speed.[193] The bees don't wait for a consensus because there isn't time. A proper balance is needed to stop paralysis by analysis.[194] The five-step process could be used successfully by any organization looking for sustainable decision-making results.

## Locust – lessons to apply

What can we glean from the decisions made by locusts? Some would say not much. Yet, both the motivation and decisions of what (to procreate) and where (in the most accessible location possible) are critical to the survival of this insect community and ours. The sheer number of locusts has impacted civilizations throughout history and continues to do so today. More is different. The fear of a locust plague is definitely a reason to pay attention if you live in one of the regions on the planet routinely impacted by these swarms.

### *Motivation.*

Perhaps it is unfair to reduce decision-making efforts of a locust to sex and appetite. Yet, the behavior of locusts after they boldly change their color, flying habit, and group gathering practices cannot be ignored and does seem to lend itself to this conclusion. Freud promoted the conscious and unconscious desire for pleasure as the foundation for human motivation. Frankl disagrees. Under the pressure cooker of a Nazi concentration camp, pleasure was a distant second. I concur that man's search for meaning, not pleasure or power, is preeminent. Yet we have this insect that provides us with an example of a community where Freud's ideas are obviously in practice. The overwhelming desire for pleasure, to be pursued in the easiest way possible, results in what we see in a locust swarm. Nothing stops the advancing horde of locusts until all is consumed and the only thing left for the locust is cannibalism. Perhaps there's a balance in our own approach to sex and appetite that should be considered. Perhaps a *mimetic* view of the word as having a given order and a given meaning and thus seeing human beings as required to discover that meaning and conform to it is in order. Of the 17 UN Sustainability Goals, the first three, no poverty, zero hunger, and good health are an excellent step in the right direction for any society. But why sustain anything? The search for purpose must be pre-eminent.

*Appetite*

Can we learn something from what the locust consumes? Is our diet that broad? Lockwood notes that the Rocky Mountain locust consumed no fewer than fifty kinds of plants from more than a dozen families (as well as leather, fabric, paper, and wool when hunger demanded), whereas the overwhelming proportion of our diet is derived from just three species—corn, wheat, and rice—found in a single family of plants.[195] Is our appetite broad enough? Do we consume a variety of vegetables? Even the consumption of locusts makes many of us cringe – yet the North American Indians included locusts as a dietary staple which, by the way has a higher protein concentration than red meat. Perhaps our decisions for sustainability should look closely at our appetite as well.

Of course, all decisions have consequences, and without understanding what to sustain and why to sustain it, the dream of how to sustain it is fraught with painful setbacks. Can we learn from these massive eusocial insect communities? These communities have an instinctual desire to survive and do. We have something more – a desire to find meaning in it all.

> *General Law – We sustain what we love, what we desire to produce, and the suffering we are willing to endure.*

 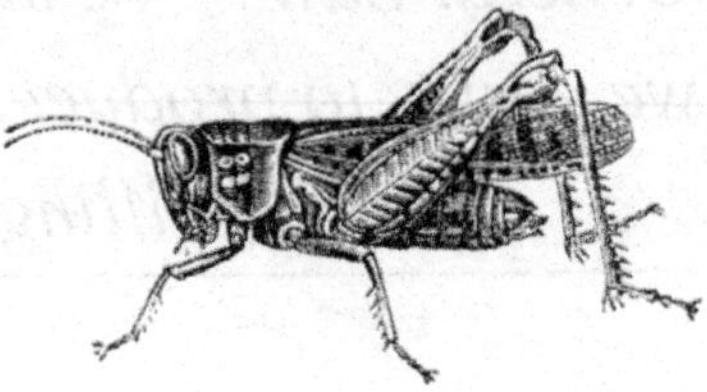

*The most painful experience of modern consciousness is the loss of center.*

**– Richard Weaver**

# Chapter 6: Communications

Goshen, Indiana, is a wonderful community with small farms and animals. While living in this part of the country, I would routinely wake up early to complete various animal chores before leaving for work. During these chores, my children were my constant companions.

We would feed chickens, turkeys, sheep, a few cows, and a horse. In the winter, various frozen watering devices would require the ice to be broken. My children would get various tools to heat the water or carry feed to the animals. Mucking out the barn was always fun and climbing into the loft just to talk added a special feel to the location.

On one occasion, my daughter was sitting next to me following our chores. The sun was shining through the window in the front porch area. It was a beautiful morning, and she looked expectantly across the farmyard at the animals. At that time, I decided to talk to her about important things that I thought had more relevance than the care of animals.

My daughter has very big green eyes and, similar to the "Love is…" cartoon famous in the '70s and '80s, could display the impression of intense interest. Therefore, I began to wax eloquently about life, commitment, and the future…

"Daddy?" she asked.

"Yes?"

"Are we having a discussion?"

"Well, uh, yes, we're having a discussion." I responded, somewhat puzzled.

"I don't like discussions, Daddy. Can we do something else, like eat?"

With that, I realized there are times when "discussions" are just not what a six-year-old wants at 7 o'clock in the morning. Breakfast is more important.

---

As a beekeeper, I have been soothed by the constant hum of a beehive during my inspections. When the bees are peaceful, they exude a pleasant noise that makes the world seem like a well-run garden—the flowers in their place and all the creatures, bees in particular, doing exactly what they are supposed to do. All are content, and all are rested. In fact, if you search for bee sounds on the internet, YouTube will provide several recorded hives that are used for therapeutic purposes. Therapeutic communication at its finest – the hum of bees.

In contrast, the beating wings of a billion flying locusts generate another emotion—fear. The sound, or lack of sound, of the extinct Rocky Mountain locust and its plundering of mid-America in the 1870s was recorded by Laura Ingalls Wilder *On the Banks of Plum Creek*:

> The light was queer. It was not like the changed light before a storm. The air did not press down as it did before a storm. Laura was frightened, and she did not know why.
>
> She ran outdoors, where Pa stood, looking up at the sky. Ma and Mary came out, too, and Pa asked, "What do you make of that, Caroline?"

> A cloud was over the sun. It was not like any cloud they had ever seen before. It was a cloud of something like snowflakes, but they were larger than snowflakes and thin and glittering. Light shone through each flickering particle.
>
> There was no wind. The grasses were still, and the hot air did not stir, but the edge of the cloud came on across the sky faster than wind. The hair stood up on Jack's neck. All at once, he made a frightful sound up at that cloud, a growl and a whine.[196]

As the locusts descended, their beating wings became louder, and the family ran furiously for shelter. Wilder writes that the sound was deafening as the locusts landed and began to consume everything in sight.

If you've read the *Little House on the Prairie* series, you may remember this event. As humans, our memories allow us to recall experiences that occurred years ago and, through the written word, even after all event participants are dead, to remember sounds. Not so with these two insect communities. They appear to need a constant input of information, repeating what has occurred. They appear to live only for the moment. It's as if their communication suffers from a strange form of amnesia.

## Clive Wearing – amnesia and communication

Clive Wearing, a former musician for the BBC, is now the most famous amnesia patient in the world. In 1985, Clive suffered a severe fever that gave him both anterograde and retrograde amnesia. That means he can neither form new memories nor recall most of his previous life. Instead, he lives his life thirty seconds at a time.[197]

Clive's struggle has been well chronicled in two documentaries, the first produced in 1986 and the second in 2005.

Clive retains some knowledge—he can play the piano expertly, for example, and remembers that he is married—yet he doesn't remember the wedding, his children, or his wife's name. The dominant experience of Clive's life, repeated hundreds of times a day, is waking up from a coma for the first time without knowing who or where he is.[198]

Though Clive is not a locust or a bee, both insects seem to have this same condition. Locust seem bent on staying together yet, like a social network, going in the same direction, telling the same story over and over again is necessary. The most frequent behavior when worker bees meet is rapidly stroking each other's antennae. The action is not grooming for stroking each other's antennae is a form of communication—remembering and transferring information about what just happened so that others may know it as well.

## Bees and communication

<u>The hum</u>

The hum of a beehive includes communication about the activities of the hive. Bees aren't communicating in any verbal way that's familiar to us; nevertheless, they are deeply engaged with each other, passing along information about the world outside and conditions within the colony. They're essentially carrying on a nonstop, colony-wide conversation like human interactions, with some exchanges quick and superficial while others are lengthy and profound.[199] Like Clive, they're repeating what is known over and over again.

So, what is the key communication issue discussed by the bees? It may sound like a political discussion, but it's more important than that!

How's our queen? She's still present, in good health, and laying eggs? Excellent. Long live the queen!

It's the most important question asked and answered in the colony because if the queen is failing or dies, the workers must begin rearing a new queen within hours, or the colony will not survive her passing. New queens are reared from young worker larvae by feeding them a special nutritious brood food called royal jelly. If queen rearing doesn't commence quickly after queen loss, the worker larvae soon become too old to switch their developmental pathway toward queens, and chaos will soon reign within the hive. When do the 30-50,000 bees know this? Within 15-30 minutes of the event.[200] The communication pathways are so well-oiled and the message so distinct (how's our queen?) that all are tuned to both the question and response.

The hive is always engaged and focused on communication. The health of the hive depends on it. Yet the care and protection of the queen, though it is critically important, is not routinely highlighted on nature channels. If you've watched any nature channel programs concerning bees, you'll be told bees are well known for the waggle dance. Humans are fascinated by the famous waggle dance. So, when did we first notice the waggle dance? Let's go back in history less than 100 years.

## The Waggle Dance for Food

Karl von Frisch, in the summer of 1944, discovered the "waggle dance." Before 1944, von Frisch thought that the only thing the dance followers learned from the dancer was the fragrance of the flowers she had visited—which bees detected by holding their antennae close to the dancer to smell the floral scents adhering to her body. From here it was thought that the bees, upon leaving the hive, simply searched in ever-expanding circles until they discovered flowers with the memorized fragrance.[201]

Yet von Frisch was gifted with the patience necessary to see more. He observed more than just the stroking of the antennae. He discovered that when a bee performs a waggle run inside a dark hive, she produces a miniaturized reenactment of her recent flight outside the hive over the sunlit countryside, and in this way, indicates the location of the rich food source she has just visited.[202] If the food source is further than 1,000 meters, the dance becomes a figure eight, and that section in the middle (the waggle) communicates two critical pieces of information – the distance to the source of pollen and the direction, in relation to the sun, from the hive to that source.

Nature programs do an excellent job communicating the magic of this dance. The simple bee waggle, observed in the confines of a dark hive, convinces and recruits others to exploit the pollen and nectar found.

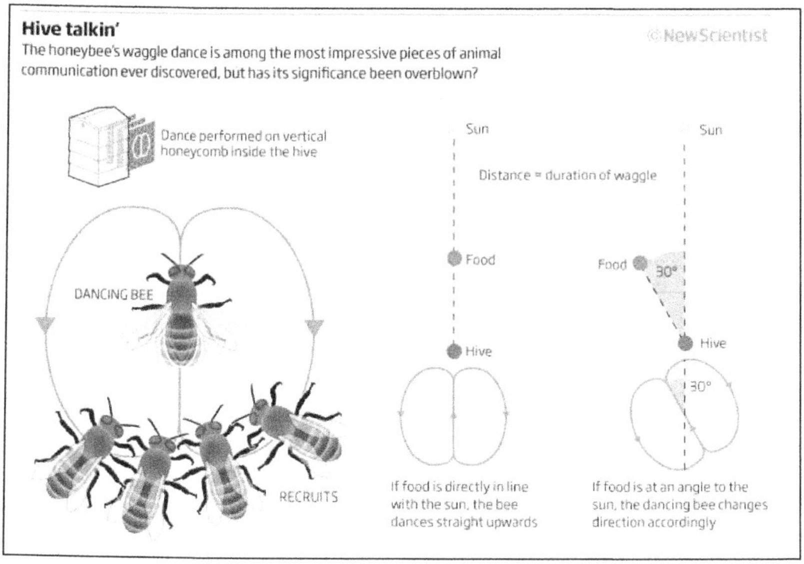


## The Waggle Dance for a New Home

The waggle dance doesn't change much in its initial usage to also locate a new home. Yet, the choice of a nest requires all the bees to acquiesce to a final decision. Unlike the search for food, where the forager is recruiting fellow food gatherers, the debate for choosing a home is a life-and-death struggle for the bees in the swarm. The quality of the future nest is critical. A dangerous home

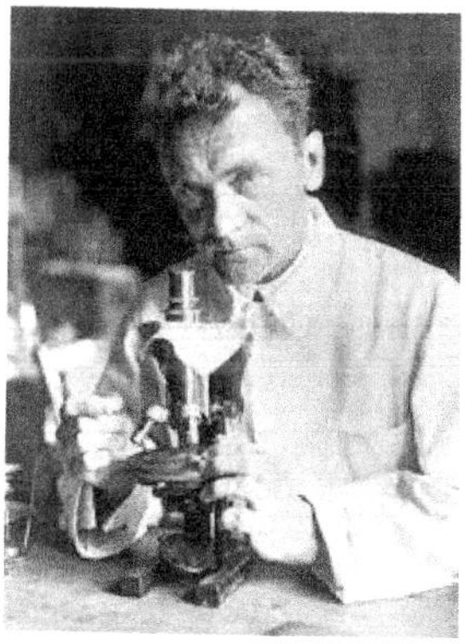

may be the end of the entire hive. In Chapter 5 we discussed the decision-making dynamics of how the scout communicates her convictions concerning finding a good nest site above the fray of others who believe theirs is the place to go. Let's look closer at the communication.

A student of Karl von Frisch, Martin Lindauer, noticed bees in a tree near the Zoological Institute in Munich doing a similar dance on the outside of a swarm.[203] The bees performing the dance were not covered in pollen, but brick dust and dirt and the information communicated was not the location of a food source but a future nest site.

Dr. Seeley of Cornell University has studied the dance and knowledge-sharing taking place in these situations. The scouts use the waggle dance to promote/advertise the potential home they discovered and use all five senses in their communication with other bees to do so. Like the waggle dance for food, the same metrics appeared to be used for the location of the proposed nest. On average, one second of the combined body-waggling and wing-buzzing represents some 1,000 meters (six-tenths of a mile) of flight. And the angle of the waggle run, relative to straight up on the vertical comb, represents the angle of the outward journey relative to the direction of the sun.[204] See the diagram on the previous page.

Communicating quality

The scout bee represents the quality of a site in two ways, first, by adjusting the rate of the dance circuit (R, in dance circuits per second) and second, by increasing the duration of the dance circuit (D, in seconds). The total number of dance circuits produced (C, in dance circuits) is the product of the rate and duration of the dancing (C = R X D).[205] Thus, the debate occurring on the outside of the ball of bees hanging from a tree branch in your yard is a novel expression of representative democracy in action, as most bees are not involved in the exploration (only 1-2% of the bees actually participate in the debate) they listen and absorb the exchange. Over a 24-72-hour period, the options decrease so that only two and eventually one possibility is considered. Once the swarm reaches a quorum and the intensity of the debate diminishes, the bees prepare to move.

## Other Methods of Bee Communication

One would expect that this visual, physical exchange of information must therefore be the primary means for honeybee communication, but it's not. Bees use all five senses to communicate, and chemical messages are perhaps the first on the list.

The "queen substance" pheromone that wafts steadily from her body is known by all within the hive. This pheromone, the major component of the material secreted by the mandibular glands in the queen's head, is a 10-carbon fatty acid whose exact name is 9-oxo-2-decenoic acid (9-ODA).[206] The queen's pheromones are a blend of at least nine compounds, all of which are necessary to fully attract worker bees. The workers transfer the information about the queen's presence or absence through these pheromones.[207]

Though the decision is not entirely the queen's, her pheromones do tell the hive to begin the development of new queens, and act to regulate genes associated with age-dependent behaviors. Her pheromones can also decrease defensive behavior by reducing guarding at the nest entrance and lowering stinging efforts following a disturbance. Concerning the energy of the hive, the pheromones act as a stimulant and can increase the tempo of comb construction. Finally, through this chemical, the colony's social hierarchy is maintained by preventing worker bees from developing their ovaries and laying eggs.[208] How does this marvelous compound 9-ODA move around the hive?

## The transfer of the queen's pheromones

The retinue bees attending the queen rub and lick off her pheromones and then move through the nest, acting as messengers proclaiming her presence. The bees then lick the antennae of other bees. The worker bees sense the queen's pheromone on contact, cleanse their palates of her odor by swallowing some of the pheromones and absorbing the rest across their cuticles, where it ends up in their bloodstream and is inactivated or performs an as-yet-unknown function. Within a relatively short period, the hive is aware of the queen and her communication.

## The Memory of a Honeybee

I mentioned that honeybees do not have a lengthy memory of certain experiences and must always be reminded of the current state of affairs. That is only true to a point. An article in *Science News*, summarized below, highlighted the memory of honeybees:

As far as the honeybee is concerned, not all flowers are created equal. Some flowers are a good source of nectar while others are a waste of time. Alfalfa flowers can kill a honeybee because they are designed for pollination by larger insects. When an insect lands, a central petal in the flower releases the stamen and sweeps pollen upward to aid pollination. Honeybees can be tossed off the flower or trapped inside by this action.

Scientists always thought that honeybees solved the problem of finding safe and productive flowers by remembering a list of characteristics. They believed the honeybee was too simple to remember mental pictures of favored flowers. However, recent experiments have convinced scientists that bees actually store a mental picture of productive flowers. In these experiments, bees were rewarded with sugar water for visiting some patterns and not others. Their reactions showed that they remember images, not simply patterns.

What's more, the experiments showed that bees are quite intelligent and good learners.[209]

Perhaps they don't have as much memory impairment as mentioned earlier.

# Locust and communication

## How do locusts communicate the message to stay together?

Locusts do not seem to communicate the need to stay together as a result of hunger or stress, but some believe because of an unbalanced aspect of fear.[210] In some mysterious locust fashion, the group cheers themselves on to stay together as they march or fly in this gregarious group. Some speculate they communicate through various pheromones that indicate the willingness of the group to eat the non-conformist. The non-conformist is easily identified as well. He or she is the one unwilling to march or fly with the advancing band. We will discuss this in future detail in the chapter addressing balance.

When not yet capable of flight (before the last instar) locusts march together. The march can be 10 miles wide and many more miles long and is conducted as one imagines the Mongol hordes of old did – as a great mass on the move from dawn to dusk with some groups or individuals within it coming to temporary halts, to feed, to rest, to bask in the sun awhile, to potter and peer, to fidget and fight, yet always to join the forward movement again.[211] Proverbs state, "The locusts have no king, yet all of them march in rank." Thus, their communication does not appear to have a centralized source but it does tell them to move and to move in ranks.

## Locust and Social Networking

In an interesting article from the Institute of Physics, Gerd Zschaler discusses a recent experiment attempting to determine why the locust move in the same direction. He concluded that

> "the mechanism through which locusts agree on a direction to move together (sometimes with devastating

> consequences, such as locust plagues) is the same we sometimes use to decide where to live or where to go out for dinner: we let ourselves be convinced by those in our social network, often by those going in the same direction but many times by those going in the opposite direction as well. We don't necessarily pay more attention to those doing the same as us, but many times we pay more attention to those doing something different."[212]

In other words, we look to the left and the right and join their ranks. If others are going in the opposite direction, we observe their advance or obstacles as well as the consequences of going against the crowd and decide whether to join ranks with the rebels or not. Locusts appear to do the same thing. Like a simple visual survey of the thumbs up and thumbs down in social media prompting human opinion to surge one way and then another, locusts make a similar appraisal of their group before deciding the next step.

Zschaler states:

> We focused on the swarming experiments of Buhl et al. In their setup, groups of 5-120 locusts were placed in a ring-shaped arena and left to march freely for 8 hours while a digital camera captured their positions and orientations. At low insect numbers, no collective order of motion arises; the system displays no clear clockwise or counterclockwise flow of locusts around the arena. At intermediate insect numbers, locusts start aligning, generating long periods of collective rotational motion during which most agents are marching in the same direction. These periods of coherent motion are interrupted by rapid spontaneous changes in their collective heading direction. Finally, at high insect numbers, this spontaneous direction switching is no longer

observed, and agents rapidly adopt a common and persistent marching direction, either clockwise or counterclockwise."[213]

Again, at low numbers, the social network doesn't work, and the locusts go where they want. At intermediate locust strength, the group begins to move uniformly in the same direction but will, at times, switch directions—almost as a group—for no apparent reason. Finally, when the numbers are large, the direction is set, and switching does not occur. The march is on. The rebels in the intermediate group ended us as little-minded conformists who joined the group late.

So, what can we understand from this? With no king (or queen), the only directional decision is from the horizontal. Unlike the bees, who look to communication from the scouts' recommendations or the queen pheromones, the locust only takes communication from what is observed through this social network. It's as if they live in a world with rooms that have only windows and no skylights. Their eyes are attuned to gather information only from the rest of the group, and the larger the group, the more conformity toward the default direction.

The parallel to our current social networking system is frightening. One of the biggest pitfalls of the thumbs up or thumbs down option in social media posts is its profound impact on how we **value ourselves.** Like the locust, once the group of opinion makers is large enough, any willingness to be an individual is lost, we cease our independence and join the group, losing the courage to remain steadfast. Many people seek external approval and validation through arbitrary likes and followers. In practice, it's no wonder why someone can suffer an existential crisis when these numbers fluctuate, and apparent "cyber-bullying" occurs. To be sustainable, one must be wary of a pressure to conform when communication and civil debate ends.

## Communication interruption

When the locust changes back to the solitary state, a major upheaval in its willingness to communicate occurs. Often, the desire of a locust reverts to that of a hermit. Its desire is solitude. It has forgotten how to listen or make eye contact. During this phase, it flies at night, attempts to blend into the surrounding by camouflage, and hides during the daytime. The goal is to avoid predators and live as much of a solitary life as possible. If locusts remained in their isolated form (looking and acting more like grasshoppers in your backyard), we wouldn't really care about their existence. But they don't.

The introverted locust in the solitary state dramatically changes when it becomes gregarious. Now the communication is face-to-face, and a threat of potential cannibalism lurks just beneath the surface.

The smartphone and social media have had a dramatic effect on human communication over the past few decades. Certain aspects of communication, such as body language, tone of voice, eye contact, and word choice, previously critical to grasping any message, have atrophied. It is as if empathy, the ability to understand the message through all the cues of the communicator, has lessened. Consequently, humans communicate more crassly, and the rudeness is palpable in tweets and news reports. When humans first enter the world of social media, the introverts are immediately empowered. The anonymity makes it easier to behave in unkind or insensitive ways without fearing the consequences. They fail to perceive others as real people with real feelings.

## Summary

### Bees – lessons to apply

The complexity of honeybee communication has fascinated modern science. The use of all the senses to exchange information allows this insect to perform amazing feats of pollination, honey production, and swarming. The search and acquisition of a new nest requires similar information exchange that far surpasses a human's ability to make optimal decisions. It can see in the dark recesses of a hive and even remember complex pattern in various flowers. With this frequent contact and obvious civil discourse, the honeybee continues to share information even in the dark.

According to brainline.org, 80-85% of our perception, learning, cognition, and activities are mediated through vision. We are visual learners for the most part and learn from watching. Our eyes once again, are critical to communication even though stress can inhibit their effectiveness.

### Locusts – lessons to apply

Similarly, the locust appears as a visual conformist whose eyes direct its path and actions as well. In the gregarious stage, it always desires togetherness. The driving desire is to see and respond to other locusts and do the same thing they are doing. The apparent move from rebel without a cause (in its solitary state) to a little-minded conformist bent on moving in the same direction as the masses demonstrates why locust swarms are so powerful and destructive and their eyes so critical for survival.

In fact, the locust maintains a unique center different than the bees. We have difficulty determining it but only because it is lost to our view. That center is an amorphous mass of conformity to self that acts as a substitute center for the swarm of locusts. They continue to fly back or walk back toward that acceptable and uniform center throughout their existence in the gregarious phase.

Thus, concerning sustainability it's not only about how they communicate but what they communicate that's important. Conformity to the center is what they seek. Similarly, in the world in which we live, it is frequently self-fulfillment that has become the new center of humanity. It is not about self-sacrifice, self-discipline, self-restraint, and self-abasement, which can be implied if not directly represented by the bee community, but something else. Perhaps by comparison, we can observe the locust pursue the meaning of it all or the center, allowing us to reorient our center to be more sustainable. Why do we do what we do? Conformity or totalitarianism? Perhaps by our wise observation of the larger story that we are a part of and making a choice.

In "A Scandal in Bohemia," Holmes instructs Watson on the difference between seeing and observing:

> "When I hear you give your reasons," I remarked, "the thing always appears to me to be so ridiculously simple that I could easily do it myself, though at each successive instance of your reasoning, I am baffled until you explain your process. And yet I believe that my eyes are as good as yours.
>
> "Quite so," he answered, lighting a cigarette, and throwing himself down into an armchair. "You see, but you do not observe. The distinction is clear. For example, you have frequently seen the steps which lead up from the hall to this room."
>
> "Frequently."

"How often?"

"Well, some hundreds of times."

"Then how many are there?"

"How many? I don't know."

"Quite so! You have not observed. And yet you have seen. That is just my point. Now, I know that there are seventeen steps, because I have both seen and observed."[214]

*General Law – To make sense of this world one must both see and observe what is being communicated.*

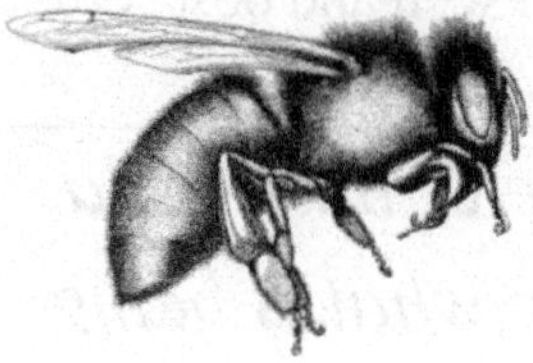

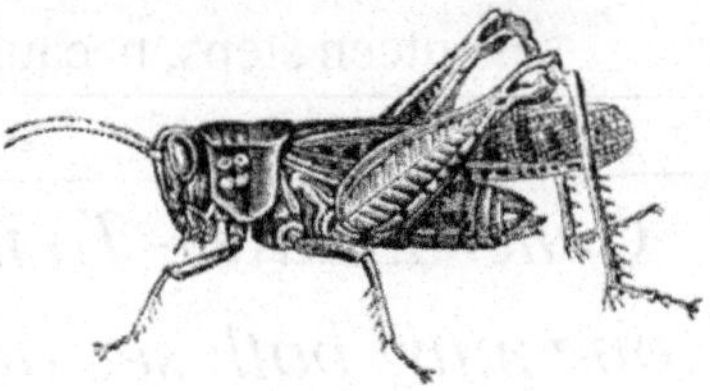

*Most people get a fair amount of fun out of their lives, but on balance, life is suffering, and only the very young or the very foolish imagine otherwise.*

– **George Orwell**

# Chapter 7: Balance and Tension

When my youngest son was 13, his fight with leukemia appeared only in the rearview mirror. Four years of chemotherapy was over and life began again. He was a happy, carefree young man willing to approach life through a new set of experiences. One of those experiences was to help raise money for the Make-a-Wish Foundation through the Trail Blaze challenge. Make-a-Wish encouraged our family during the trying times of leukemia and he wanted to repay their kindness.

The Trail Blaze challenge is a 26-mile hike requiring participants to raise at least $2,000 from sponsors in order to participate. Once the financial support goal is met, one must begin training for the hike. My son was the youngest participant in 2018, so he (as well as his father) needed to spend some quality time preparing.

To hike 26 miles in 12 hours, one needs the right equipment and the right training. The equipment was easy. Hiking shoes and socks are available at most sporting goods stores. Camel-pack day packs with trail-mix aren't a problem. All we needed was the experience and muscles to complete the challenge, and we would be balanced.

The preparation hikes started early in the morning. First, a few miles and then a few more. We used two hiking sticks to keep us from falling and worked on speed. Eventually, we moved from 4 to 6 to 10-mile hikes at a 3.5 to 4 mile per hour pace. We were ready.

The scheduled hike started with our group meeting at 4 AM and driving to a nearby state park. My son and I were excited, and the energy of beginning the challenge with about 60 other participants was in the air. At 6 AM, the organizers fired a gun and we were off.

The first 10 miles weren't that bad. We maintained our pace and talked with other hikers. When we got to the lunch-break location at about the 14-mile mark, we were starting to feel the strain of the hike. Eventually, our speed dropped, the rain began to fall, and we found ourselves near the end of the group.

We kept pushing ourselves and almost made it to the finish. Yet the balance of both equipment AND practice became apparent. We needed more experience to meet the challenge the next time. Yet the tension of choosing the hike, the sweat, and the contest will never be forgotten. The struggle to pursue a worthwhile goal bore fruit. Our relationship as father and son grew stronger that day.

---

Frankl mentions that mental health is based on a certain degree of tension. The tension between what one has already achieved and what one still ought to accomplish or the gap between what one is and what one should become.

> I consider it a dangerous misconception of mental hygiene to assume that what man needs in the first place is equilibrium or, as it is called in biology, "homeostasis." i.e., a tensionless state. What man actually needs is not a tensionless state but rather the striving and struggling for a worthwhile goal, a freely chosen task.[215]

Similarly, balance in a sustainable society will always include tension. It may be a trite and overused quote but the only constant in life is change, and any failure to grasp that concept will result in confusing sustainability with homeostasis. Balance and tension must go hand in hand.

Consider the arch again. If an architect wants to strengthen a decrepit arch, he increases the load which is laid upon it and thereby the parts are joined more firmly together and strengthened. Societies that struggle and succeed at balance and tension are those that will be sustainable. Those struggles will be called 'stressors' by some. Let's look at the stressors which confront bees and locusts.

## Bees and maintaining balance

Honeybees, like us, are challenged to balance their role as individuals relative to the communal good. All societies are constantly on a continuum between the personal and the collective, but bees are very much in favor of their community. Although we may or may not decide to emulate their socially conscience efforts, their teamwork is impressive and something to remember. How do the honeybees manage the physical, environmental, and task-oriented stress that is part of being a successful community of insects? Let's start with temperature.

### Physical stressors

In the heat of the summer, some Langstroth boxes are routinely covered with what appears to be more honeybees on the outside of the boxes then inside. The bees are so thick that the hive looks almost black from a distance. As a result of the heat and the humidity, the bees must take action. Acting as a single organism, they work at controlling the temperature of their home. From late winter to early fall, when the workers are rearing brood, a colony's internal temperature is kept between 34° and 36 ° C (93 ° - 96 ° F) – just below the core body temperature of humans – even as the ambient air temperature ranges from -30 ° C to 50 ° C (-20 ° to 120 ° F) temperature. Maintenance of this temperature is crucial. As a colony each individual bee accomplishes this by adjusting the rate at which it sheds or generates heat. The honeybee does this by resting its metabolism during high temperature periods and, in times of extreme cold, by boosting its metabolism to intensify heat production.[216] In the heat of July, the majority of the bees in my backyard were on the outside of the hives to minimize the temperature impact their bodies would have on the brood and queen inside. Some were still inside, doused with water and fanning their wings to cool the hive to the necessary temperature but most felt it necessary to maintain the balance by leaving.

<u>*Managing heat within the hive*</u>

Heat is generated within the cluster by the bees isometrically contracting their two sets of flight muscles (one for elevating the wings and one for depressing them), thereby producing much heat but little or no wing vibration. These flight muscles endow a bee with a surprisingly powerful engine. Bees fly, of course, by flapping their wings (the most energetically demanding mode of animal locomotion) and using their flight muscles. These muscles are among the most metabolically active of all their tissues and excellent for generating heat.[217] Of course, with heat generation comes $CO_2$.

<u>*Managing $CO_2$, circulation, and fevers in the hive*</u>

The honeybees also regulated their *colonial breathing* – limiting the buildup of the respiratory gas $CO_2$ inside the hive by increasing its ventilation. When the $CO_2$ level reaches 1-2% of the air in the nest cavity the bees begin this practice. Furthermore, the hive also has *colonial circulation* – keeping the heat-producing bees in the central, brood-nest region of the hive properly fueled with honey carried in from peripheral honeycombs. Finally the bees can imitate a *colonial fever response* – mounting a disease-fighting elevation of the nest temperature when a dangerous fungal infection of the brood bees occurs.[218] Like our human body, the bees can fight the infection by raising the temperature of the hive.

Similarly, human civilization flourishes when the management of waste, wastewater, and air is paramount. The

History Channel aired a program entitled *Engineering an Empire* from 2005 to 2007. The empires highlighted were able to sustain their civilizations by successfully addressing physical stressors such as water, waste, and wastewater prior to any military conquest. A tension exists between our basic human functions and the possible diseases spread from being too close together. To sustain one must never forget this important if not critical infrastructure requirement. When you flush the toilet, where does the waste go? When 10 million people in a city like New York do the same, how is all that human excrement managed? Only a sustainable society has thought that through and kept these logistical issues at the forefront of their plans.

Environmental stressors

Environmental stressors are those that occur outside the hive but can have a detrimental impact on its very operation. Some of these will surprise you.

*Robber bees*

Honeybees are compulsive hoarders. They collect nectar or honey from any source they can find, and that includes a poorly guarded or weak hive. Rusty Burlew, in an article on robbing bees in *Honeybee Suite*, describes it well:

> Think of it like this: It is a hot August afternoon. It hasn't rained in weeks. The flowers are long past their peak, and the few that remain are crispy. A gang of bored workers with too much time and not enough to do is hanging out, looking for trouble. Suddenly, one of the gang picks up on a scent…sweet! It's coming from a nearby hive where the beekeeper has spilled some syrup. A few scouts check it out and believe they can overpower the lethargic guard bees lounging in the heat. Within minutes the dancers post directions on CombBook, and the siege is on![219]

The fight results in the inclusion of wasps who are attracted to the dead bees like vultures. The robbing bees are very noisy and approach the hive swaying from side to side. Soon the robbing bees lose their hair and appear shiny and black because of the fighting. Eventually the battle ends, and the weaker hive is decimated. Dead bees are everywhere. Honeycomb cells are open, pieces of wax litter the scene. Death and destruction is all that remains.

The sustainability lesson is obvious – the best defense is a good offense. If the guard bees had been less lethargic and if the warning system (pheromones) quicker in notifying the hive, the smaller hive may have survived. A sustainable community can never forget this tension of greed for resources. Altruistic activities do take place among nation states, but a quid pro quo assumption lurks in the background. Like bees, we must understand this tendency to survive and sustain a community. If not, the ignorance of failing to prepare for robbing nations will sink our best efforts in building something to last.

*Industrial pressures (both hives and billionaire beekeepers)*

Nearly 2 million hives are transported across the U.S. every year to pollinate the nation's farmlands. Pollinator companies drive trucks packed with beehives from one side of the country to the other. As mentioned earlier, the effort begins in the almond fields of California every spring and spreads across the country from southwest to northeast. Most of these farms are mono-cultural (i.e., one crop in current rotation), but there are actually three types of farms—organic, conventional, or intermediate, and industrial. A lack of available pollinators could impact all of these. The availability of readily accessible pollinators does not seem to be a stressor of concern for sustainability – or does it?

A student of Mark Winston in Canada performed an economic analysis showing that yield and profit could be maximized if about 30% of organized farmland were left uncultivated. Farmers who planted their entire field would earn approximately $27,000 in profit per farm, whereas those who left 33% unmanaged would earn around $65,000 in profits due to improved pollination and the resultant increase in yield. These are

startling as well as counterintuitive results: up to a point, planting less acreage resulted in higher yields and greater income simply by relying on unmanaged land to increase wild bee populations.[220] Why? Winston details the novelty of a diverse pollination effort that included both the industrialized (and delivered) honeybees and the local bee and pollinator population. By providing the two pollination groups the opportunity to work together, since they appear to have different methods of operation (discussed earlier), a smaller farm produced a more fruitful harvest.

Sustainable communities in the past have taken similar lessons to heart. Rotating fields with diverse crops to ensure key macronutrients are not depleted, or not harvesting the corner sections of a farm, or allowing certain fields to remain fallow every seven years are all well-known sustainable methods for maximizing the harvest for the long-term. A sustainable community would continue looking for this balance to maximize short-term outputs with the tension to keep the farm and output healthy.

Task stressors

A final group of stressors that upset the balance and tension of working together as an insect community are those associated with routine activities necessary for a functioning hive.

*Honey storage and usage*

When the hive is in full operation, the comb consists of thousands of back-to-back hexagonal cells, engineered to remarkable precision, within 0.05 mm tolerance in diameter and 0.002 mm in cell wall thickness.[221]

These cells are used for raising brood (larvae and developing bees), pollen, and honey. If you were to slice the hive in sections, the center comb would be populated by brood followed in an outward circle (like layers of an onion) of pollen and eventually of honey. An unbalanced hive that positions the important center of the hive too close to the perimeter risks losing brood to seasonal temperature variations. A beginning beekeeper is told to ensure the brood is near the center of the box.

Similarly, we find the protection of the children in a sustainable community in the center, protected from the unknown. Like the emergence of young bees from the brood area whose wings are too soft to take flight, the protection of children from the elements is key to sustaining the next generation. Exposure of children, like the exposure of young bees to an environment they are unprepared to face, can spell disaster.

A tension exists on this continuum – when do you expose the young bees and children to their eventual destiny? Dory in the movie *Finding Nemo* explains the tension well when confronting an overprotective father: "You can't never let anything happen to him. Then nothing would ever happen to him." Eventually parents in a sustainable society must find the balance and tension in this decision to protect and a desire to release their beloved children. Searching for this balance consumes so much of our lives. In the past 75 years even the Chinese government, understanding and confusing the importance, has voiced a concern. First a one-child policy and now a multiple child policy. Is this a government decision (the *poietic* choice) or a family decision (the *mimetic* choice)?

*Too Many Drones*

If a hive goes queenless for any period, workers with more developed ovaries may begin laying eggs. Since these workers lack the male sperm collected by the queen during her mating flight, they can only lay drones. A hive consists of 99+% female workers. For the beehive, the drones are an odd addition to the team and do not promote much to the daily functioning of the hive. Their importance is passing on the genetic strength of one hive to another when cavorting with the young queens on their mating flight. It seems almost tragic that a worker, desiring to continue the health of the hive, can only produce bees whose final development will result in eating all the contents of the stored pollen and honey in the hive. The unbalanced hive, without the queen, is now careening toward absolute destruction as a result of the good intentions of these workers. A worker can't drone and a drone can't work.

Perhaps this analogy is going too far for some but the one demographic most prevalent in the prison population is absent fathers. The instability of the most basic unit, the family causes chaos in society. For example, children brought up in single mother homes are:

5 times more likely to commit suicide,
9 times more likely to drop out of high school,

10 times more likely to abuse chemical substances,
14 times more likely to commit rape,
20 times more likely to end up in prison, and
32 times more likely to run away from home.[222]

A familiar saying that 'moms can't dad and dads can't mom' would seem to be the lesson here.

A sustainable community must see the *mimetic* balance in this affair. A master plan for raising children is a delicately balanced system. These well-intentioned workers who take up the task of laying eggs only wanted to see the hive continue. Their efforts were noble and sacrificial. Yet, they failed. The hive or the sustainable community collapses.

## Locust and maintaining balance

### Are locust outbreaks evidence of a disturbed and out of balance system?

We seem to believe that, as with a well-behaved child or a good worker, all species should refrain from extreme outbursts. The concept of a Golden Mean or "everything in moderation" is rooted in the ancient Greeks, the progenitors of Western culture. This idealization of the world has been variously manifested throughout history. We anticipate little to no change. Gradualism is the name of the game.[223] No catastrophes. No global shifts in the day-to-day events of life. But if a few billion locusts fly or march over the horizon and begin eating every plant in sight, one believes we must have a disturbed system. Unfortunately, this is not true.

Only the activities of the Food and Agriculture Organization of the United Nations have provided a reprieve over the past 75 years. These 75 years have been a period of relative calm from something that is quite routine from a historical viewpoint. The desert locusts, in the gregarious state, are the species with the most economic impact of any insect in the world and its cycle of existence has been sustained for a very, very long time. Alone it threatens one-fifth of Earth's land area and one-tenth of the global population. Over 60 countries are susceptible to swarms from the desert locust species alone.

So, if the swarming locust is not a disturbed system, why does its size vary so dramatically and so quickly? There is something about locusts that once learned can make us exceedingly wise. Perhaps through the locust we can see the needed balance and tension of a truly sustainable community that has tremendous adaptability.

## Physical stressors

### *Starvation and Thirst*

For a locust the length of time for food to pass through the digestive system is between 30 minutes and a couple of hours, but when food supplies are short the digestive system of a locust can dramatically change.[224] The desert locust's protruding eyes, will scan the arid, lonely land at times and find no sign of green. As a result of what it sees, in some mysterious way, the rate at which it converts food will be retarded so that it may take from three to four days (not hours) to pass through its body. During this time, at least, it will be able to resist starvation. Based on its vision, the locust can slow down its digestive tract to less than 1-3% of its prior pace.

Not only can the locusts survive without food, the amount of water in its body can drop by approximately half yet still enable it to search for a new source of supply.[225]

Because of the tension of existing in a desert environment, these physical adaptations can only make one marvel. Every continent on earth presents a temperature, water, or food acquisition challenge. To be sustainable, one must consider the limitations and advantages we must have to maintain balance. A tension for precious resources like food and water will continue in any sustainable community. Adaptability will be the key.

According to VOANews, 40% of the water consumed in 2021 in Singapore (a city of 5.6 million) came from wastewater sources. Could this practice occur elsewhere?[226]

## Environmental stressors

### *Predators (eating toxic plants)*

The locust is near the bottom of the food chain, so predators are many. The stress of staying alive for this insect is palpable. Its eyes are always in search of something desiring to make the locust the next meal.

Locusts are highly adaptable insects capable of consuming a wide range of plant material. In fact, their ability to feed on a variety of plants is one of the reasons why they are such successful and destructive pests. By eating toxic plants, they can deter, to some degree, the predators that are in search of an easy lunch.

By expanding its diet, the locust does two things, it increases available food sources and decreases its attractiveness as an easy snack. Both steps can be recognized as improvements in the opportunity for the insect to sustain itself in a hostile environment. Over 80% of our human diet comes from three species – corn, wheat, and rice – found in a single family of plants. A sustainable community would need to expand this.

*Cannibalism*

A frightening physical stressor faced by locust is cannibalism. A study suggests that cannibalistic interactions among individuals and the threat of attack by those approaching from behind is a principal factor in the onset of collective movement among locusts. Thus, their apparent willingness to march in rank is not a longing to join the military but a fear of being bitten by a neighboring locust. Fear from being bit from behind, individual locusts increase the propensity of others to march as they approach toward or contact their abdomen. Those that don't move face an increased risk of cannibalism.[227]

Thus, the forward momentum of the marching bands or flying swarms is prompted by a tension we can barely understand. The locust moves so as not to be consumed. A balance for maintaining their individual's life in the gregarious phase must include almost constant forward motion.

What would you see as a lesson from such a situation? Perhaps don't be complacent about those in your vicinity. A tension concerning personal safety must always be a consideration.

## Summary

Sustainability addresses the existence of a certain degree of tension when seeking for balance. Locusts must contend with a department of the UN organized to control and if possible, destroy the swarms that haunt the deserts of our planet. Honeybees, though excellent role models for diligence, must contend with the possibility of collapse, attack from other hives, and over eager workers that want to make things right by stepping into the position of surrogate queen while lacking the necessary abilities to do so.

Viktor Frankl knows this better than most, but the deaths in the 20th century because of war is staggering. According to The School of Public Policy at the University of Maryland, 231 million people were killed as a result of war in the 20th century.[228] In contrast, approximately 70 million people died as a result of famines during the same period. Man is not basically peaceful and good at heart. Our very nature made us the instrument of of three times as much death from war as from famines.

Homeostasis, or a tensionless state for the insect communities does not exist. Similarly, what man actually needs in a sustainable community is not a tensionless state either but rather the commitment to strive and struggle for a worthwhile goal, a freely chosen task as well as the sobering fact that we tend to cause our own destruction. Accepting this reality and the tension that exists is a significant step toward sustainability.

> *General Law – Maturity is achieved when a person accepts life as full of tension.*
>
> *Joshua L. Liebman*

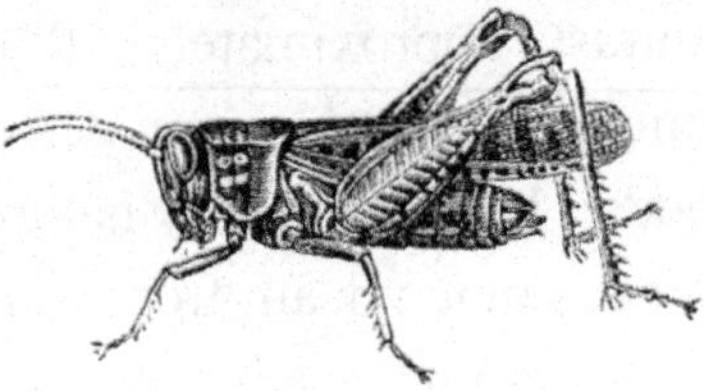

*There must be more to life than having everything!*

– **Maurice Sendak**

# Chapter 8: The Search for the Story

I drive an old Toyota Prius to and from work. The car is perhaps the most successful hybrid yet built and was certainly popular in the early part of the 21st century. Why? Perhaps the consumption screen. The consumption screen is the default display on the dash and tracks the amount of fuel consumed per mile traveled in 5-minute increments. If you are unaware of the screen, it was a

singularly successful way Toyota used to visually communicate one's effort on being "more sustainable" – using less to move from point A to point B. As an engineer in a pharmaceutical plant, looking at trend charts of energy usage is routine. Yet, by affixing the bar graph prominently on the dashboard of the car, it motivates me to see if I can continue the trend upward in my travels with every passing mile. To use less gas to travel the same distance. To be, in some way, more sustainable at this five-minute interval than in the past.

I applaud all those that compete on this plane of reducing their footprint, of using less to accomplish more, but sustainability, and hopefully this book, is showing you that sustainability is something a lot more then efficiency. Like the search for meaning, the search for being more sustainable must address the why along with the how.

---------------------------------------------------

The Importance of Stories

Through the art of storytelling, we preserve our heritage, educate future generations, and inspire the search for meaning. Stories have a way of bringing to the forefront those elements of the situations or facts that seems to excite our brain in new and fascinating ways. According to various sources (not all of them medically accurate), a story activates parts in the brain that allow the listener to turn the story into their own ideas and experiences.

Listeners and readers not only experience similar brain activity to each other, but also to the author or speaker. As a result, the brain releases dopamines into the system when it experiences an emotionally charged event, making it easier to remember and with greater accuracy.

As noted in the opening chapter, our society is listening to stories, but not trusting them. We go to the movies and watch television and place those stories into our own individualistic design, missing the perspective of the true order that is around us.

## *Mimesis* vs. *Poiesis*

Aristotle helped frame the debate about viewing the story of the world in terms of perspective. A *mimetic* view regards the world as having a given order and a given meaning and thus sees human beings as required to discover that meaning and conform themselves to it. *Poiesis*, by way of contrast, sees the world as full of raw material out of which meaning, and purpose can be created by the individual.

Western society has moved away from the *mimetic* view of the world as possessing intrinsic meaning to a different and, I believe dangerous view where we create meaning along the way. As a result, we find ourselves in an existential vacuum. If what we see around us is not to educate us on how to fit into the grand story of all things, what are we to do but to breathe and die?

## The Existential Vacuum and Sunday Neurosis

Frankl discussed the existential vacuum in his book *Man's Search for Meaning*. Existential because it addresses our existence at this moment in time, and vacuum because there is a continual void with the human question of why we exist at all. Unlike bees and locusts, we don't have an instinctual drive for performance and life. Boredom tends to inhabit our continually growing amount of free time as we contemplate the puzzle – why are we here?

Frankl mentions the Sunday neurosis – that kind of depression which afflicts people who become aware of the lack of contentment in their lives when the rush of the busy week is over, and the void within themselves becomes manifest usually during the weekends. Not a few cases of suicide can be traced back to this existential vacuum. Depression, aggression, and addiction seem to be apparent symptoms.

Like the locust waiting for the wind, the modern world thinks horizontally—searching for a non-existent center among each other—and the loss of centeredness prompts the vacuum we contend with to grow in power. The vacuum expands as our thoughts constantly change, and insofar as they are only our best thoughts on the subject at any one time, they are always subject to continual revision.

In contrast honeybees do have a center (metaphorically) and seek to protect that center (the queen) through their existence. As a community, the message of what needs to be sustained for the bees is clear, and the work is distinct. The community prepares for the worst winter through the healthy and prosperous spring and summer season. It then braces for the changing weather and begins the cycle again.

Yet for those in an existential vacuum, we operate more like the locusts who only think horizontally. Aware of the pressures on all sides from other locusts, the locust keeps its mandated social distance (except for the drive to copulate) always moving forward with this inane idea that someone must know where they are going. The vacuum grows stronger with each passing day as all that can be consumed is consumed, and the cannibalistic tendencies of the insect set in. They continue to wait for the wind and march to the center but without a center (metaphorically) in existence except more consumption.

What are we to learn from this? Using a *mimetic* view, one regards the world as having a given order and a given meaning and thus sees human beings as required to discover that meaning and conform themselves to it. The *poietic* view will lead humanity into a spiral of continual frustration. I hope you see the world differently as a result of these insect communities – each doing its best to sustain the larger group as it prepares for the next season of life.

## The meaning of life and the meaning of sustainability

Frankl mentions a situation posed to a chess champion when asked to answer the question, "Tell me, Master, what is the best move in the world?"

> There simply is no such thing as the best or even a good move apart from a particular situation in a game and the particular personality of one's opponent—similarly, the same holds for human existence. One should not search for an abstract meaning of life. Everyone has his own specific vocation or mission in life to carry out a concrete assignment which demands fulfillment. Therein he cannot be replaced, nor can his life be repeated. [229]

Like the honeybee and locust, the question of what sustainability is cannot be answered as a singular abstract concept. Sustainability is a normative term, describing and valuing what is of worth at that moment. Each person must find, through searching what they desire to sustain. The community is sustained when each discovers that meaning and conforms their live to what they discover. Thus, each of us is asked, "What is the best move for me to take to help sustain the family, community, or the world?" It's different and unique for each of us. – yet it is not created, it is found.

## A painter or an ophthalmologist

I hope this presentation of the honeybee and locust did more than just paint a picture. I desire not to be a painter but an ophthalmologist to enable you to see the world as it really is. These two insect communities exist in the world today. One is critical to the food supply. The other is desirous of devouring that food supply before it sees the end of another day. Hopefully you can see lessons not previously understood from these communities.

Scholars say that a single specific definition of sustainability may never be possible. But the concept is useful, and as a normative term, it communicates the value we place on things. We want to sustain this, and we don't want to sustain that. But the essence of sustainability requires us to ask why sustain anything? Unlike the two insect communities, as humans a value judgment will move us to the question of why? And that question of why will lead us to answer with – love.

## The Meaning of Love

The locust swarm did demonstrate rampant sexual activity, but this is not love. Sex may be an outward demonstration of love to a degree, but sex demonstrated by neither insect community exhibited the love of sacrifice. Even though John Chrystom wrote poetically that "The bee is more honorable than other animals not because she labors, but because she labors for others" that instinctual labor was not a labor of love. Labor for others like the sacrifice of a soldier, police, and 343 fireman that entered the World Trade Center on 9/11 are better examples. Frankl mentions in the concentration camp how the willingness of one individual to give his life to another by sharing a piece of bread, a smile, or the attitude to not despair brought the light of life's meaning to the prisoners. Through this, they were willing to sustain themselves for another day.

## The Meaning of Suffering

The story of sustainability on various media platforms is presented as a survival scenario. One's willingness to be part of helping society survive is to suffer. The images paint the scenario as a hopeless dystopian novel in which only the elimination of humanity (the "hard" sustainability approach) can save the planet.

Frankl faced a real threat to his daily existence in the Nazi concentration camps. How did he address the suffering and his desire to survive? He did so through attitude.

> The moment suffering ceases to be suffering is when it finds meaning. Life has a meaning up to the last moment, and it retains this meaning literally to the end. Life's meaning is an unconditional one. Meaning precedes happenstance.[230]

Meaning or what this life is all about and why we are here to sustain anything precedes happenstance.

During one particularly harrowing situation where Frankl's odds of survival were twenty-eight to one, most of his comrades in the concentration camp posed the question, "Will we survive the camp? For, if not, all this suffering has no meaning." Frankl found the despair in the question a critical indicator of the problem. The question which beset Frankl was, "Has all this suffering, this dying around us, a meaning? For, if not, then ultimately there is no meaning to survival; for a life whose meaning depends upon such a happenstance – as whether one escapes or not – ultimately would not be worth living at all."[231] Meaning precedes happenstance, and our survival and sustainability on this planet follow why we are here in the first place.

## Life's transitoriness

Concerning sustainability, many national or international goals and changes are decades, if not centuries, into the future. Yet the transitoriness of our existence in no way makes our impact meaningless. Our existence requires of us responsibleness. Some approach the duty flippantly – "Let the next generation worry about sustaining the (fill in the blank) issue." But let us not be pessimistic but rather activistic. Learn from those societies around us – "go to the ant, you sluggard, observe her ways and be wise." Go to the beehive and go to the locust swarm. Learn from these communities and their practices.

## Anticipatory Anxiety

The choice of these two insect communities, the beehive and locust swarm, is a way of my addressing anticipatory anxiety. These two natural societies lacking the human "why" are caught in the web of our existential vacuum and demonstrate two fearful approaches to environmental issues – the totalitarian approach (beehives) and the conformist approach (the locust swarm). Animal instincts result in decisions and methods to cope with how to address the environments that are part of these creature's DNA. Yet we are more than animals. Blaise Pascal, concerning humanity in *Pensée* states it best –

> Man is but a reed, the most feeble thing in nature; but he is a thinking reed. The entire universe need not arm itself to crush him. A vapour, a drop of water suffices to kill him. But, if the universe were to crush him, man would still be more noble than that which killed him, because he knows that he dies and the advantage which the universe has over him; the universe knows nothing of this.[232]

Pascal brilliantly articulated humanity's value, as well as our complex relationship with the rest of the world. As we rabidly pursue how to be sustainable without a "why" to sustain anything at all, we grow in anxiety and frustration.

C.S. Lewis once wrote: "Mistaken for our mother, [nature] is terrifying and even abominable. But if she is only our sister—if she and we have a common Creator—if she is our sparring partner—then the situation is quite tolerable."[233] Understanding this may calm our anxious hearts as we contemplate sustainability.

## So what do we do with this anxiety?

Man is more than a machine, and his continued sustainability on this planet is more than just an engineering problem of how to harness the amount of energy needed and produce the water and food to continue.

Lord Moulton, a jurist in England – proposed that for any society to survive, to be healthy and productive, three domains must be preserved: the domain of law, the domain of freedom, and the internal domain of 'obedience to the unenforceable.' This third domain requires a search for the "why" in life. The obedience to the unenforceable is a search beyond engineering and challenges each of us in the search to become more sustainable. For as Nietzsche stated, "He who has a why to live for can bear almost any how."

## Why did the attack occur on September 11, 2001?

Most Western think tanks and the general population at large has settled the issue – 9/11 is the emergency call number established in 1968 and was chosen by the suicide bombers for shock effect. Why else would anything happen on that particular date?

Christopher Hitchens suggested that we may not be considering the history of that date as clearly as we should. He stated:

I now think I can provide a more persuasive explanation, however. It was on September 11, 1683, that the conquering armies of Islam were met, held, and thrown back at the gates of Vienna.

Now this, of course, is not a date that has only obscure or sectarian significance. It can rightly, if tritely, be called a hinge-event in human history. The Ottoman empire never recovered from the defeat; from then on it was more likely that Christian or western powers would dominate the Muslim world than the other way around. In our culture, the episode is often forgotten or downplayed. But in the Islamic world, and especially among the extremists, it is remembered as a humiliation in itself and a prelude to later ones. (The forces of the Islamic Jihad in Gaza once published a statement saying that they could not be satisfied until all of Spanish Andalusia had been restored to the faithful as well.)[234]

For nearly two months the city of Vienna had been under siege. A large Ottoman army of more than 150,000 soldiers surrounds the city. Their commander is Grand Vizier Kara Mustafa. The Ottoman Empire had attempted to conquer the city 150 years earlier and failed. Kara Mustafa is back at the gateway to the Holy Roman Empire to try again.

In the months leading up to Kara Mustafa's attack, the city had done what it could to quickly improve its fortifications. It also sent delegates to surrounding states appealing for armed assistance. As the Ottoman Empire forces drew near, many of Vienna's citizens fled.... leaving only about 11,000 trained soldiers and 5,000 volunteers behind to defend the city and wait for reinforcements.

Grand Vizier Kara Mustafa loses many men in the 2-month siege, but by September 10th, his diminished—yet still substantial army—is just hours away from breaching the wall and gaining access to the city. Relief forces have begun arriving and take up positions on the high ground less than 10 miles away…they are awaiting the arrival of Polish King John III Sobieski and his 3,000-member cavalry known as the "Winged Hussars." Their delay means there's a large gap in the line.

Kara Mustafa makes a strategic error that not only loses the battle but marks the beginning of the end of the Ottoman Empire's dominance in Europe. Historian Matthias Pfaffenblicher spoke with the Discovery Channel in 2016.

He didn't concentrate all his troops in *defending his camp against the attacking relief army but continued the assault of the city because it was so close for the fall. So he divided his force.*[235]

When Sobieski and his mercenary cavalry do show up, they rout the Ottoman detachment. The Winged Hussars along with 15,000 other horse mounted forces throw themselves upon the rest of Mustafa's army. It is the largest known cavalry charge in Europe's history. Kara Mustafa and his army flee the field. Vienna is saved.

The battle marks the first time a largely united Europe cooperates militarily against the Ottomans. Grand Vizier Kara Mustafa chooses not to counterattack. He returns home in shame where he is executed and beheaded.[236]

Was it history? Relevant or irrelevant?

# Appendix A

## Albert's Swarm

If you live in the United States, you are probably quite satisfied that you've never had to address locusts. It's something that happens elsewhere – but not in North America. Well, are you prepared for a surprise? Where was the largest Locust Swarm ever observed? In Nebraska. Albert Child, a doctor who wanted to be a meteorologist, moved to Nebraska in 1857. He began sending reports about weather patterns to the Smithsonian Institution in 1861. Albert Child came to be the right man, in the right place, at the right time to provide a definitive report of the most immense swarm of locusts in recorded history. He began his account with the objective tone of a trained and experienced observer:

> The extent of the swarm is difficult to ascertain, as the observer can only see a small belt. They may extend indefinitely right or left. During the flight from June 15 to 25 of 1875, I telegraphed east and west. I found a continuous line moving northward of 110 miles and then somewhat broken 40 miles further. The movements of the winds for five days (15th to 20th) averaged about 10 miles per hour, and the locust evidently moved considerably faster than the wind, at least 15 miles per hour.[237]

Child provided the width, speed, and length of time the Rocky Mountain Locust flew overhead, but how thick was the swarm? The child was not only a meteorologist and doctor but a very resourceful one as well. He used a telescope on various objects of similar size to a locust placed at known distances across the rolling hills. He then used the focus of his telescope to conclude "at least as high on the horizontal focal distance that he'd just determined" to estimate the thickness of the swarm.

> The swarm I estimated from one-quarter to one-half mile deep. It seemed like piercing the milky way of the heavens; my glass found no limits to them. They might have been a mile or more in-depth.[238]

So, with these estimates, Dr. Child presented his straightforward calculations of the aggregate size of the locust swarm that was passing over Nebraska. He was incredulous about his own findings:

> They were visible from six to seven hours of each of the successive five days, and I can see no reason to suppose that their flight was checked during the whole five days. If so, the army in the line of advance would be 120 hours by 15 miles per hour = 1,800 miles in length, and say at even 110 miles in width, an area of 198,000 square miles! And then from one-quarter to one-half mile deep. This is utterly incredible, yet how can we put it aside?

An area of 198,000 square miles would encompass the combined areas of Connecticut, Delaware, Maine, Maryland, Massachusetts, New Hampshire, New Jersey, New York, Pennsylvania, Rhode Island, and Vermont.[239] This is one swarm – perhaps numbering more than 3.5 trillion locusts at one time!

That number may surprise you. Dr. Lockwood puts it into perspective.

Now let's compare the biomass of the Rocky Mountain Locust to the Bison – the roaming herds of 30 to 100 million that crossed the plain states before 1850. Let's assume the 1870 locust outbreak was normal and use half of the estimated number. Thus, locusts would routinely occupy an estimated 500,000 square miles (about twice the area of Texas). The carrying capacity of rangeland for modern-day grasshoppers is around 10 individuals per square yard. Using this figure, which is probably quite conservative, as we're considering the locust during an outbreak—we'd end up with 15 trillion insects, or a couple of thousand locusts for every person currently on the earth.[240]

Now let's compare the biomass of the Rocky Mountain Locust and the Bison. Why?
Ecologists tend to prefer biomass—the number of kilograms, pounds, or tons of living tissue for comparison. An average bison weighed 500 pounds (35 pounds for a newborn calf and a ton for a mature male). If there were 45 million bison in 1870, then there were somewhere around 11 million tons of critters scattered over the continent. As for the locusts, let's assume a weight of half a gram per locust. Given 15 trillion insects, their collective biomass would have been about 8.5 million tons. So, bison and locusts had similar and rather phenomenal masses of herbivorous tissue. And both were very nearly gone by the turn of the nineteenth century.[241]

During this same period, after a swarm had departed, one Minnesota farmer went into his fields to see how many eggs had been left behind… turning over a shovelful of soil and making a careful count, he found that there were 150 eggs per square inch. "At that rate, there will be 940,896,000 eggs to the acre or the nice little pile of 6,586,272,000 on seven acres of my farm." Similarly, in Utah, there were 118 eggs per square inch.[242]

How were these large swarms addressed? The eggs left behind by the swarms were the one life stage against which the settlers could readily launch a counterattack. The key was to alter the amount of water and disturb the protective cover of the soil. Females preferred to oviposit in well-drained sandy soil, so the eggs were normally kept dry by virtue of their parents' having carefully chosen a suitable nursery. With irrigation, the farmers could inundate their fields in the fall or, more commonly, in the spring before the eggs hatched. Next came the plow, which could disturb the egg pods deposited approximately 6" below the soil's surface. Next came the ditches. Through trial and error, the entomologists discovered that ditches could halt the marching locusts across the landscape. Over time, the group discovered the optimal ditch dimensions: eighteen inches wide and two feet deep—too wide for the nymphs to hop across and too deep for them to escape.[243] Once trapped, the nymphs could be drowned or burned.

Yet, the Rocky Mountain Locust is no longer in existence – either destroyed by the practices put in place to control their numbers or the discovery and eradication of their primary breeding ground – no one is certain of their demise.

# Appendix B

## Life Change Index Scale (The Stress Test)

| Event | Impact | My Score |
|---|---|---|
| Death of spouse | 100 | |
| Divorce | 73 | |
| Marital Separation | 65 | |
| Jail Term | 63 | |
| Death of close family member | 63 | |
| Personal injury or illness | 53 | |
| Marriage | 50 | |
| Fired at work | 47 | |
| Marital reconciliation | 45 | |
| Retirement | 45 | |
| Change in health of family member | 44 | |
| Pregnancy | 40 | |
| Sex difficulties | 39 | |
| Gain of a new family member | 39 | |
| Business readjustment | 39 | |
| Change in financial state | 38 | |
| Death of a close friend | 37 | |
| Change to a different line of work | 36 | |
| Change in number of arguments with spouse | 35 | |
| Mortgage over $200,000 | 31 | |
| Foreclosure of mortgage or loan | 30 | |

| | | |
|---|---|---|
| Change in responsibilities at work | 29 | |
| Son or daughter leaving home | 29 | |
| Trouble with in-laws | 29 | |
| Outstanding personal achievement | 28 | |
| Spouse begins or stops work | 26 | |
| Begin or end school | 26 | |
| Change in living conditions | 25 | |
| Revisions of personal habits | 24 | |
| Trouble with boss | 23 | |
| Change in work hours or conditions | 20 | |
| Change in residence | 20 | |
| Change in schools | 20 | |
| Change in recreations | 19 | |
| Change in church activities | 19 | |
| Change in social activities | 19 | |
| Mortgage or loan less than $200,000 | 17 | |
| Change in sleeping habits | 16 | |
| Change in number of family get-togethers | 15 | |
| Change in eating habits | 15 | |
| Vacation | 13 | |
| Christmas approaching | 12 | |
| Minor violation of the law | 11 | |
| Total | | |

**Directions: If an event mentioned above has occurred in the past year, or is expected in the near future, copy the number in the score column. If the event has occurred or is expected**

**to occur more than once, multiply this number by the frequency of the event.**

**Scoring The Life Change Index**

The body is a finely timed instrument that does not like surprises. Any sudden change stimuli which affect the body, or the reordering of important routines that the body become used to, can cause needless stress, throwing your whole physical being into turmoil.

The following chart will give you some idea of how to informally score yourself on Social Readjustment Scale. Since being healthy is the optimum state you want to achieve, being sick is the state of being you most want to avoid.

| Life Change Units | Likelihood Of Illness In Near Future |
|---|---|
| 300+ | about 80% |
| 150-299 | about 50% |
| less than 150 | about 30% |

The higher your life change score, the harder you have to work to get yourself back into a state of good health.

T.H. Holmes and T.H. Rahe. "The Social Readjustment Rating Scale," Journal of Psychosomatic Research. 11:213, 1967.

# Appendix C

## The Volunteer at Auschwitz

Between 1940 and 1945, as many as two million people were murdered at the Nazi concentration camp at Auschwitz and the neighboring extermination site of Treblinka in south-central Poland. Countless acts of courage and faith took place amid the horror there. Viktor Frankl mentions one name in his book *Man's Search for Meaning* as an example of someone who had obviously found meaning – Maximilian Kolbe. This is his story.

Maximilian Kolbe was forty-five years old in the early autumn of 1939 when the Nazis invaded his homeland. He was a Polish friar in Niepokalanow, a village near Warsaw. There, 762 priests and lay brothers lived in the largest friary in the world. Father Kolbe presided over Niepokalanow with a combination of industry, joy, love, and humor that made him beloved by the plainspoken brethren there.

In his simple room, he sat each morning at a pigeonhole desk, a large globe before him, praying over the world. He did so, tortured by the fact that a pale man with arresting blue eyes and a terrifying power of manipulation had whipped the people of Germany into a frenzy. Whole nations had already fallen to the evil Adolf Hitler and his Nazis.

"An atrocious conflict is brewing," Father Kolbe told a group of friars one day after he had finished prayers. "We do not know what will develop. In our beloved Poland, we must expect the worst." Father Kolbe was right. His country was next.

On September 1, 1939, the Nazi blitzkrieg broke over Poland. After several weeks, a group of Germans arrived at Niepokalanow on motorcycles and arrested Father Kolbe and all but two of his friars who had remained behind. They were loaded on trucks, then into livestock wagons, and two days later arrived at Amtitz, a prison camp.

Conditions were horrible, but nor horrific. Prisoners were hungry, but no one died of starvation. Strangely, within a few weeks the brothers were released from prison. Back at the friary, they found the buildings vandalized and the Nazis in control, using the facility as a deportation camp for political prisoners, refugees, and Jews.

The situation was an opportunity for ministry, and Father Kolbe took advantage of it, helping the sick and comforting the fearful.

While Kolbe and the friars used their time to serve others, the Nazis used theirs to decide just how to impose their will on the rest of Europe. To Adolf Hitler, the Jews and Slavic people were the *Untermenschen* (sub humans). Their cultures and cities were to be erased and their industry appropriated for Germany. On October 2, Hitler outlined a secret memorandum to Hans Frank, the governor general of Poland. In a few phrases he determined the grim outcome for millions: "The ordinary Poles are especially born for low labor. The Polish gentry must cease to exist. All representatives of one master for the Poles, the German."

As for Poland's hundreds of thousands of priests?

"They will preach what we want them to preach," said Hitler's memo. "If any priest acts differently, we will make short work of him. The task of the priest is to keep the Poles quiet, stupid, and dull-witted."

Maximilian Kolbe was clearly a priest who "acted differently," from the Nazis' designs.

In early February 1941, the Polish underground smuggled word to Kolbe that his name was on a Gestapo list: he was about to be arrested. Kolbe knew what happened to loved ones of those who tried to elude the Nazis' grasp: their friends and colleagues were taken instead. He had no wife or children; his church was his family. And he could not risk the loss of any of his brothers in Christ. So he stayed at Niepokalanow.

At nine o'clock on the morning of February 17, Father Kolbe was sitting at his pigeonhole desk, his eyes and prayers on the globe before him, when he heard the sound of heavy vehicles outside the thick panes of his green-painted windows. He knew it was the Nazis, but he remained at his desk. He would wait for them to come to him.

After being held in Nazi prisons for several months, Father Kolbe was found guilty of the crime of publishing unapproved materials and sentenced to Auschwitz. Upon his arrival at the camp in May 1941, an SS officer informed him that the life expectancy of priests there was about a month. Kolbe was assigned to the timber detail; he was to carry felled tree trunks from one place to another. Guards stood by to ensure that the exhausted prisoners did so at a quick trot.

Years of slim rations and overwork at Niepokalanow had already weakened Kolbe. Now, under the load of wood, he staggered and collapsed. Officers converged on him, kicking him with their shiny leather boots and beating him with their whips. He was stretched out on a pile of wood, dealt fifty lashes, then shoved into a ditch, covered with branches, and left for dead.

Later, having been picked up by some brave prisoners, he awoke in a camp hospital bed alongside several other near-dead inmates. There, miraculously, he revived.

"No need to waste gas or a bullet on that one," chuckled one SS officer to another. "He'll be dead soon."

Kolbe was switched to other work and transferred to Barracks 14, where he continued to minister to his fellow prisoners, so tortured by hunger they could not sleep.

By the end of July 1941, Auschwitz was working like a well-organized killing machine, and the Nazis congratulated themselves on their efficiency. The camp's five chimneys never stopped smoking. The stench was terrible, but the results were excellent: eight thousand Jews could be stripped, their possessions appropriated for the Reich, gassed, and cremated – all in twenty-four hours. Every twenty-four hours.

About the only problem was the occasional prisoner from the work side of the camp who would figure out a way to escape. When these escapees were caught, as they usually were, they would be hanged with special nooses that slowly choked out their miserable lives – a grave warning to others who might be tempted to try.

Then one July night as the frogs and insects in the marshy land surrounding the camp began their evening chorus, the air was suddenly filled with the baying of dogs, the curses of soldiers, and the roar of motorcycles. A man had escaped from Barracks 14.

The next morning there was a peculiar tension as the ranks of phantom-thin prisoners lined up for morning roll call in the central square, their eyes on the large gallows before them. But there was no condemned man standing there, his hands bound behind him, his face bloodied from blows and dog bites. That meant the prisoner had made it out of Auschwitz. And that meant death for some of those who remained.

After the roll call, Camp Commandant Fritsch ordered the dismissal of all but Barracks 14. While the rest of the camp went about its duties, the prisoners from Barracks 14 stood motionless in line. They waited. Hours passed. The summer sun beat down. Some fainted and were dragged away. Some swayed in place but held on; those the SS officers beat with the butts of their guns. Father Kolbe, by some miracle, stayed on his feet, his posture as straight as his resolve.

By evening roll call the commandant was ready to levy sentence. The other prisoners had returned from their day of slave labor; now he could make a lesson out of the fate of this miserable barracks.

Fritsch began to speak, the veins in his thick neck standing out with rage. "The fugitive has not been found," he screamed. "Ten of you will die for him in the starvation bunker. Next time, twenty will be condemned."

The rows of exhausted prisoners began to sway as they heard the sentence. The guards let them; terror was part of their punishment.

The starvation bunker! Anything was better – death on the gallows, a bullet in the head at the Wall of Death, or even the gas in the chambers. All those were quick, even humane, compared to Nazi starvation, for they denied you water as well as food.

The prisoners had heard the stories from the starvation bunker in the basement of Barracks 11. They said the condemned didn't even look like human beings after a day or two. They frightened even the guards. Their throats turned to paper, their brains turned to fire, their intestines dried up and shriveled like desiccated worms.

Commandant Fritsch walked the rows of prisoners. When he stopped before a man, he would command in bad Polish, "Open your mouth! Put out your tongue! Show your teeth!" And so he went, choosing victims like horses.

His dreary assistant, Palitsch, followed behind. As Fritsch chose a man, Palitsch noted the number stamped on the prisoner's filthy shirt. The Nazis, as always, were methodical. Soon there were ten men – ten numbers neatly listed on the death roll. The chosen groaned, sweating with fear. "My poor wife!" one man cried. "My poor children! What will they do?"

"Take off your shoes!" the commandant barked at the ten men. This was one of his rituals; they must march to their deaths barefoot. A pile of twenty wooden clogs made a small heap at the front of the grassy square.

Suddenly there was a commotion in the ranks. A prisoner had broken out of line, calling for the commandant. It was unheard of to leave the ranks, let alone address a Nazi office; it was cause for execution.

Fritsch had his hand on his revolver, as did the officers behind him. But he broke precedent. Instead of shooting the prisoner, he shouted at him.

"Halt! What does this Polish pig want of me?"

The prisoners gasped. It was their beloved Father Kolbe, the priest who shared his last crust, who comforted the dying and nourished their souls. Not Father Kolbe! The frail priest spoke softly, even calmly, to the Nazi butcher. "I would like to die in place of one of the men you condemned."

Fritsch stared at the prisoner, No. 16670. He never considered them as individuals; they were just a gray blur. But he looked now. No. 16670 didn't appear to be insane.

"Why?" snapped the commandant.

Father Kolbe sensed the need for exacting diplomacy. The Nazis never reversed an order; so he must not seem to be asking him to do so. Kolbe knew the Nazi dictum of destruction: the weak and the elderly first. He would play on this well-ingrained principle.

"I am an old man, sir, and good for nothing. My life will serve no purpose."

His ploy triggered the response Kolbe wanted. "In whose place do you want to die?" asked Fritsch.

"For that one," Kolbe responded, pointing to the weeping prisoner who had bemoaned his wife and children.

Fritsch glanced at the weeping prisoner. He did look stronger than this tattered No. 16670 before him.

For the first and last time, the commandant looked Kolbe in the eye. "Who are you?" he asked.

The prisoner looked back at him, a strange fire in his dark eyes. "I am a priest."

"*Ein Pfaffe*!" the commandant snorted. He looked at his assistant and nodded. Palitsch drew a line through No. 5659 and wrote down No. 16670. Kolbe's place on the death ledger was set.

Father Kolbe bent down to take off his clogs, then joined the group to be marched to Barracks 11. As he did so, No. 5659 passed by him at a distance – and on the man's face was an expression so astonished that it had not yet become gratitude.

But Kolbe wasn't looking for gratitude. If he was to lay down his life for another, the fulfillment had to be in the act of obedience itself. The joy must be found in submitting his small will to the will of One more grand.

As the condemned men entered Barracks 11, guards roughly pushed them down the stairs to the basement.

"Remove your clothes!" shouted an officer. Christ died on the cross naked, Father Kolbe thought as he took off his pants and thin shirt. It is fitting that I suffer as He suffered.

In the basement the ten men were herded into a dark, windowless cell.

"You will dry up like tulips," sneered one jailer. Then he swung the heavy door shut.

As the hours and days passed, however, the camp became aware of something extraordinary happening in the death cell. Past prisoners had spent their dying days howling, attacking one another, clawing the walls in a frenzy of despair.

But now, coming from the death box, those outside heard the faint sounds of singing. For this time the prisoners had a shepherd to gently lead them through the shadows of the valley of death, pointing them to the Great Shepherd. And perhaps for that reason Father Kolbe was the last to die.

On August 14, 1941, there were four prisoners still alive in the bunker, and it was needed for new occupants. A German doctor named Boch descended the steps of Barracks 11, four syringes in his hand. Several SS troopers and a prisoner named Brono Borgowiec (who survived Auschwitz) were with him – the former to observe and the latter to carry out the bodies.

When they swung the bunker door open, there, in the light of their flashlight, they saw Father Maximilian Kolbe, a living skeleton, propped against one wall. His head was inclined a bit to the left. He had the ghost of a smile on his lips and his eyes wide open, fixed on some faraway vision. He did not move.

The other three prisoners were on the floor, unconscious but alive. The doctor took care of them first, a jab of the needle into the bony left arm, the push of the piston in the syringe. It seemed a waste of the drug, but he had his orders. Then he approached No. 16670 and repeated the action.

In a moment, Father Kolbe was dead.[244]

# Bibliography

Anstey, Michael L., Stephen Rogers M., R. Ott Swidbert, Malcolm Burrows, and Stephen J. Simpson. "Serotonin Mediates Behavioral Gregarization Underlying Swarm Formation in Desert Locust." *Science*, January 30, 2009.

Augustyn, Adam, The Editors of Encyclopedia Britannica, ed. "Locust." In *Encyclopedia Britannica*, January 13, 2023. https://www.britannica.com/animal/locust-insect.

Avery, Gayle, and Harald Bergsteiner. *Sustainable Leadership: Honeybee and Locust Approaches*. New York and London: Routledge, 2011.

Baron, Stanley. *The Desert Locust*. New York: Charles Scribner, 1972.

Bazazi, Sepideh, Jerome Buhl, Joseph Hale, Michael L. Anstey, Gregory A Sword, Stephen Simpson, and Iain D Couzin. "Collective Motion and Cannibalism in Locust Migratory Bands." *Current Biology*, May 20, 2008.

Bell, Simon, and Stephen Morse. *Sustainability Indicators: Measuring the Immeasurable?* London: Earthscan, 2004.

Bennett, William J., The Volunteer at Auschwitz by Chuck Colson. *The Book of Virtues*. Simon and Schuster: New York, 1993.

Britannica, T. Editors of Encyclopedia Britannica, ed. "Locust." In *Encyclopedia Britannica*, January 13, 2023. https://www.britannica.com/animal/locust-insect.

Burlew, Rusty. "How to recognize robbing bees and ways to stop them." Accessed September 19, 2023. https://www.honeybeesuite.com/robbing-bees-questions-and-answers/

Chaudhry, Inayatullah, Sayed El Bashir, and Hassan Ali. "Sexual Behavior and Communication in the Desert Locust, Schistocerca Gregaria (Orthoptera: Acrididae): Sex Pheromone in Solitaria." *Environmental Entomology*, December 1, 1994. https://doi.org/10.1093/ee/23.6.1544.

Cheng, Ken. "How Honeybees Find a Place: Lessons from a Simple Mind." *Animal Learning & Behavior*, February 2000.

Conan Doyle, Sir Arthur. *Adventures of Sherlock Holmes*. Book of the Month Club: New York, 1994.

Cooper, Lane. *Louis Agassiz as a Teacher Illustrative Extracts on His Method of Instruction*. gutenberg.org: Project Gutenberg - e-book, 2004.

Ehrlich, Paul. *The Population Bomb*, Ballantine Book: New York, 1971.

EPA. "Colony Collapse Disorder." EPR—Pollinator Protection, October 26, 2022. https://www.epa.gov/pollinator-protection/colony-collapse-disorder.

FAO Staff. "Locust Watch: Desert Locust." *Food and Agricultural Organization of the United Nations*, 1994. https://www.fao.org/ag/locusts/en/info/info/faq/index.html
.

Frisch, Karl von. "The Dance Language and Orientation of Bees." Web Post—Bee Health, August 20, 2019. https://bee-health.extension.org/dance-language-of-the-honey-bee/#:~:text=Direction%20of%20the%20food%20source,the%20direction%20of%20the%20sun.

Gladwell, Malcolm. *The Tipping Point: How Little Things Can Make a Big Difference*. Boston: Little Brown, 2000.

Golov, Y, J Rillich, M Douek, A.R. Harari, and A Ayali. "Sexual Behavior of the Desert Locust during Intra-and Inter-Phase Interactions." *Journal of Insect Behavior*, November 21, 2018. https://link.springer.com/article/10.1007/s10905-018-9703-z.

Hackett, Kevin, and Mary Purcell-Miramontes. "CCD Colony Collapse Disorder Progress Report." Washington, D.C.: US Department of Agriculture, June 2010. https://www.ars.usda.gov/is/br/ccd/ccdprogressreport2010.pdf.

Hitchens, Christopher. "Why the suicide killers chose September 11." *The Guardian*. October 3, 2001.

Holmes, T.H., and R.H. Rahe. "The Social Readjustment Rating Scale." *Journal of Psychosomatic Research*, 1967.

Horn, Tammy. *Bees in America: How the Honey Bee Shaped a Nation*. Lexington, KY: The University Press of Kentucky, 2006.

Hooper, Walter, editor. *Present Concerns: Essays by C.S. Lewis*. New York: Harcourt Brace Jovanovich, 1986.

Irving, Justin A., and Karin Klenke. "Telos, Chronos, and Hermēneia: The Role of Metanarrative in Leadership Effectiveness through the Production of Meaning." *International Journal of Qualitative Methods*, September 2004. https://doi.org/10.1177/160940690400300303.

Kirby, William, and William Spence. *An Introduction to Entomology (Gutenberg Project)*. Fifth. London: Longman and Brackman, n.d.

Leander, Kasey, and John Stonestreet. "What Love Really Looks Like: The Man With No Memory." *BreakPoint Commentary*. August 12, 2022.

Lockwood, Jeffrey A. *Locusts: The Devastating Rise and Mysterious Disappearance of the Insect That Shaped the American Frontier*. New York: Basic Books, 2004.

Maggiacomo, Taylor, and Kaya Lee Berne. "How to Set off a Plague of Locusts." *National Geographic*, January 16, 2020. https://www.nationalgeographic.com/magazine/graphics/how-to-set-off-a-plague-of-locusts-interactive?loggedin=true&rnd=1675525373610.

Morris, William, ed. *The American Heritage Dictionary of the English Language*. Boston, MA: Houghton Mifflin Company, 1979.

Novotney, Amy. "Dangerous Distraction." *Monitor on Psychology. American Psychological Association*. Volume 40, No. 2. February 2009, page 32.

Page, Robert E. Jr. *The Spirit of the Hive: The Mechanisms of Social Evolution*. Cambridge, MA: Harvard University Press, 2013.

Pascal, Blaise. *Pascal's Pensées.* New York: E. P. Dutton & Co., Inc. 1958. Available on https://www.gutenberg.org/files/18269/18269-h/18269-h.htm#SECTION_I

Physics, Institute of. "Swarms of Locusts Use Social Networking to Communicate." *Science Daily.Com*, July 15, 2011. www.sciencedaily.com/releases/2011/07/110714191435.htm.

Predel, Reinhard, and Susanne Neupert. "Social Behavior and the Evolution of Neuropeptide Genes: Lessons from the Honeybee Genome." *BioEssays*, May 2007.

Rushkoff, Douglass. *Present Shock: When Everything Happens Now*. New York: Penguin Group, 2013.

Seeley, Thomas D. *Honeybee Democracy*. Princeton, NJ: Princeton University Press, 2010.

Shufelt, Jim, ed. *Bugle Notes*. 1st ed. Vol. 72. 72 vols. United States Military Academy, West Point, New York: USMA, 1980.

Staff Writer. "To honeybees, a picture is worth a thousand line angles." *Science News*. Volume 127. No. 13. March 13, 1985.

Staff Writer, "King Tut's Sweet Tooth." *Office for Science and Society Newsletter*, Montreal: Mcgill University Press, May 31, 2017. https://www.mcgill.ca/oss/article/did-you-know-history/king-tuts-sweet-tooth.

Surowiecki, James. *The Wisdom of Crowds: Why the Many Are Smarter Than the Few and How Collective Wisdom Shapes Business, Economies, Societies and Nations*. New York: Doubleday, 2004.

Symmons, P.M., and K. Cressmon. Desert Locust Guidelines: Biology and Behavior. Rome: Food and Agriculture Organization of the United Nations, 2001.

Tarpy, David. “Beginner Beekeeping Training.” NCSU Beekeeper Education and Engagement System, 2021.

Toffler, Alvin. *Future Shock.* New York: Bantam, 1984.

Frankl, Viktor. *Man's Search for Meaning*. 1st ed. Boston, MA: Beacon Press, 2006.

Weigle, Kent. “Crowdsourcing: Using Humanities Greatest Asset.” *2022*, n.d.

Wilder, Laura Ingalls. *On the Banks of Plum Creek.* New York: HarperCollins, 1993.

Winston, Mark. *Bee Time: Lessons from the Hive*. Cambridge, MA: Harvard University Press, 2014.

Zimmer, Carl. “The Secret Lives of Bees.” *Smithsonian Magazine* (blog), March 2012. https://www.smithsonianmag.com/science-nature/the-secret-life-of-bees-99559587/.

# Index

## U



## V



## W



## Y



## Z

## ENDNOTES

[1] Viktor Frankl, *Man's Search for Meaning*, (Boston, MA: Beacon Press, 2006), 111.
[2] Alvin Toffler, *Future Shock*, (New York: Bantam, 1984), 13.
[3] Douglass Rushkoff, *Present Shock: When Everything Happens Now*. (New York: Penguin Group, 2013), 13.
[4] Justin A. Irving and Karin Klenke, "Telos, Chronos, and Hermēneia: The Role of Metanarrative in Leadership Effectiveness through the Production of Meaning," *International Journal of Qualitative Methods*, https://doi.org/10.1177/160940690400300303, (September 2004), 30.
[5] Ibid., 6.
[6] Frankl, *Man's Search for Meaning*, 99.
[7] Ibid., 99.
[8] Rushkoff, *Present Shock*, 19.
[9] Ibid., 38.
[10] Ibid., 39.
[11] Ibid., 23. - Aristotle, quoted from Robert McKee, *Story; Substance, Structure, Style, and the Principles of Screenwriting*, (New York: Regan Books, 1997).
[12] Ibid., 48.
[13] Carl R. Trueman, *The Rise and Triumph of the Modern Self* (Wheaton, IL: Crossway, 2020), 39.
[14] Ibid., 39.
[15] Ibid., 40–41.
[16] Mark Winston, Bee Time: Lessons from the Hive (Cambridge, MA: Harvard University Press, 2014), 150.
[17] Eugene H. Merrill, and Rooker, Mark F., eds., *The World and the Word: An Introduction to the Old Testament,* (Nashville, TN: B&H Publishing, 2011), 64.
[18] Ibid., 66.
[19] Rushkoff, *Present Shock*, 232.
[20] *Lane Cooper, Louis Agassiz as a Teacher Illustrative Extracts on His Method of Instruction* (gutenberg.org: Project Gutenberg - e-book, 2004), 10..
[21] Simon Bell and Stephen Morse, *Sustainability Indicators: Measuring the Immeasurable?* (London: Earthscan, 2004), Introduction.
[22] Frankl, *Man's Search for Meaning*, 107.
[23] Ibid., 109.
[24] George Bernard Shaw, *Man and Superman: A Comedy and a Philosophy*, (The Project Gutenberg Ebook; 2015), https://www.gutenberg.org/files/3328/3328-h/3328-h.htm.
[25] Winston, *Bee Time*, 6.
[26] Ibid., 9.

[27] Thomas D. Seeley, *Honeybee Democracy*, (Princeton, NJ: Princeton University Press, 2010), 20.
[28] Winston, *Bee Time*, 6.
[29] Ibid., 21.
[30] Ibid., 21.
[31] Ibid., 25 - Dara Moskowitz Grumdahl, *Minneapolis City Pages*.
[32] Staff Writer, "King Tut's Sweet Tooth," Office for Science and Society Newsletter, (Montreal: Mcgill University Press, May 31, 2017), https://www.mcgill.ca/oss/article/did-you-know-history/king-tuts-sweet-tooth.
[33] Winston, *Bee Time*, 11.
[34] Ibid., 12.
[35] Ibid., 10.
[36] Ibid., 22.
[37] Seeley, *Honeybee Democracy*, 31.
[38] Winston, *Bee Time*, 27.
[39] Ibid., 87.
[40] November 27th, 1948, address at the Vatican - made by Pius XII to beekeepers on the lessons of bees for mankind.
[41] Winston, *Bee Time*, 12.
[42] Ibid., 13.
[43] Ibid., 194.
[44] Ibid., 13.
[45] Seeley, *Honeybee Democracy*, 23.
[46] Ibid., 23.
[47] Winston, *Bee Time,* 155.
[48] Ibid., 156.
[49] Ibid., 200.
[50] Ibid., 186.
[51] EPA, "Colony Collapse Disorder" (*EPR - Pollinator Protection,* October 26, 2022), https://www.epa.gov/pollinator-protection/colony-collapse-disorder.
[52] Ibid., 1.
[53] Ibid., 1.
[54] FAO Staff, "Locust Watch: Desert Locust," Food and Agricultural Organization of the United Nations, 1994, https://www.fao.org/ag/locusts/en/info/info/faq/index.html.
[55] William Kirby and William Spence, *An Introduction to Entomology* (Gutenberg Project), Fifth Edition, (London: Longman and Brackman, n.d.), 215–16.
[56] *The Bible (ESV)*, Exodus 10:12-13.
[57] *The Bible (ESV)*. Revelations 9:7, Joel 1:6
[58] Stanley Baron, *The Desert Locust* (New York: Charles Scribner, 1972), 3.
[59] Kirby and Spence, *An Introduction to Entomology (Gutenberg Project)*, 216.
[60] Baron, *The Desert Locust,* 115.

[61] FAO Staff, FAQ.
[62] Jeffrey A. Lockwood, *Locusts: The Devastating Rise and Mysterious Disappearance of the Insect That Shaped the American Frontier* (New York: Basic Books, 2004), 146.
[63] Baron, *The Desert Locust,* 46.
[64]Ibid., 46.
[65] Lockwood, *Locusts*, 148.
[66] Ibid., 146.
[67] Adam Augustyn, The Editors of Encyclopaedia Britannica Britannica, ed., "Locust," in *Encyclopedia Britannica*, January 13, 2023, https://www.britannica.com/animal/locust-insect.
[68] P. M. Symmons and K. Cressmon. *Desert Locust Guidelines: Biology and Behavior.* (Rome: Food and Agriculture Organization of the United Nations, 2001), 2.
[69] Baron, *The Desert Locust*, 9.
[70] Lockwood, *Locusts*, 231.
[71] Symmons. *Desert Locust Guidelines*, 4.
[72] Taylor Maggiacomo and Kaya Lee Berne, "How to Set off a Plague of Locusts," *National Geographic*, January 16, 2020, https://www.nationalgeographic.com/magazine/graphics/how-to-set-off-a-plague-of-locusts-interactive?loggedin=true&rnd=1675525373610.
[73] Maggiacomo, *National Geographic,* graphic..
[74] Lockwood, *Locusts*, 27.
[75] " FAO Staff, FAQ.
[76] Symmons. *Desert Locust Guidelines*, 6.
[77] Lockwood, *Locusts*, 21.
[78] Ibid., 146.
[79] Baron, *The Desert Locust*, 44.
[80] Lockwood, *Locusts*, 20.
[81] FAO Staff, FAQ.
[82] Ibid.
[83] Ibid.
[84] Baron, *The Desert Locust,* 47.
[85] FAO Staff, FAQ.
[86] Ibid.
[87] Ibid."
[88] Baron, *The Desert Locust*, 42.
[89] Ibid., 42.
[90] FAO Staff, FAQ.
[91] Ibid.
[92] Symmons, *Desert Locust Guidelines*, 36.
[93] T.H. Holmes and R.H. Rahe, "The Social Readjustment Rating Scale," (*Journal of Psychosomatic Research*, 1967), 214.

[94] AR Sutin et al., "Differential Personality Change Earlier and Later in the Coronavirus Pandemic in a Longitudinal Sample of Adults in the United States," (*Plos One*, September 28, 2022), https://doi.org/10.1371/journal.pone.0274542.
[95] Winston, Bee Time, 61.
[96] Kevin Hackett and Mary Purcell-Miramontes, "CCD Colony Collapse Disorder Progress Report" (Washington, D.C.: US Department of Agriculture, June 2010), https://www.ars.usda.gov/is/br/ccd/ccdprogressreport2010.pdf.
[97] EPA Staff, "Colony Collapse Disorder" (EPR - Pollinator Protection, October 26, 2022), https://www.epa.gov/pollinator-protection/colony-collapse-disorder.
[98] Winston, *Bee Time*, 71.
[99] Ibid., 71.
[100] Ibid., 71.
[101] Ibid., 72.
[102] Ibid., 80.
[103] Tammy Horn, Bees in America: How the Honey Bee Shaped a Nation, (Lexington, KY: The University Press of Kentucky, 2006), 38.
[104] Winston, *Bee Time*, 141.
[105] Robert E. Page Jr., The Spirit of the Hive: The Mechanisms of Social Evolution, (Cambridge, MA: Harvard University Press, 2013), 10-13.
[106] Winston, *Bee Time,* 184.
[107] Ehrlich, Paul. *The Population Bomb*, (Ballantine Book: New York, 1971), 25.
[108] Winston, *Bee Time,* 194.
[109] Jim Suhfelt, editor. *Bugle Notes*. 1st edition. Vol. 72. 72 vols. (United States Military Academy, West Point, New York: USMA, 1980), 147.
[110] David Tarpy, Basic Beekeeping.Class provided by NCSU Beekeeper Education and Engagement System. September 2021.
[111] Seeley, *Honeybee Democracy,* 35.
[112] Ibid., 84.
[113] Gayle Avery and Harald Bergsteiner, *Sustainable Leadership: Honeybee and Locust Approaches* (New York and London: Routledge, 2011), 73.
[114] Avery, *Sustainable Leadership*, 167.
[115] Winston, *Bee Time*, 151.
[116] Avery, *Sustainable Leaderships*, 44.
[117] Lockwood, *Locusts*, 22.
[118] Baron, *The Desert Locust,* 32.

[120] Ibid., 36.

[121] Ibid., 102.
[122] Ibid., 75.
[123] "FAO Staff, FAQ.

[124] Baron, *The Desert Locust,* 30.
[125] Lockwood, *Locusts*, 147.
[126] Michael L. Anstey et al., "Serotonin Mediates Behavioral Gregarization Underlying Swarm Formation in Desert Locust," (Science, January 30, 2009), Abtract.
[127] Baron, *The Desert Locust,* 31.
[128] Lockwood, *Locusts*, 147.
[129] Frankl, *Man's Search for Meaning*, 104.
[130] Frankl, *Man's Search for Meaning*, 105.
[131] https://knowyourphrase.com/as-busy-as-a-bee
[132] Winston, *Bee Time*, 174.
[133] Ibid.,, 178.
[134] Ibid., 175.
[135] Amy Novotney, "Dangerous Distraction." (*Monitor on Psychology*. American Psychological Association. Volume 40, No. 2. February 2009),32.
[136] Ibid., 179.
[137] Rushkoff, *Present Shock,* 56–57.
[138] Winston, *Bee Time*, 95.
[139] Ibid., 95.
[140] Ibid., 108.
[141] Ibid., 115.
[142] Ibid., 116.
[143] Avery, *Sustainable Leadership*, 83.
[144] Seeley, *Honeybee Democracy*, 42.
[145] Ziglar, Born to Win, need page number.
[146] Winston, *Bee Time*, 181.
[147] Seeley, *Honeybee Democracy*, 5.
[148] Ibid., 35–36.
[149] Baron, *The Desert Locust*, 29.
[150] Ibid., 34.
[151] Ibid., 117.
[152] Ibid., 117.
[153] Ibid., 118.
[154] Y Golov et al., "Sexual Behavior of the Desert Locust during Intra-and Inter-Phase Interactions," (Journal of Insect Behavior, November 21, 2018), https://link.springer.com/article/10.1007/s10905-018-9703-z.
[155] Lockwood, *Locusts*, 146.
[156] Rushkoff, *Present Shock*, 203-4.
[157] Seeley, *Honeybee Democracy*, 7.
[158] Winston, *Bee Time*, 184.
[159] Karl von Frisch, "The Dance Language and Orientation of Bees," (Web Post - Bee Health, August 20, 2019), https://bee-health.extension.org/dance-language-of-the-honey-

bee/#:~:text=Direction%20of%20the%20food%20source,the%20direction%20of%20the%20sun.
[160] Winston, *Bee Time*, 187.
[161] Seeley, *Honeybee Democracy*, 56.
[162] Ibid., 91–93.
[163] Ibid., 206.
[164] Winston, *Bee Time*, 188.
[165] Seeley, *Honeybee Democracy*, 67.
[166] James Surowiecki, *The Wisdom of Crowds: Why the Many Are Smarter Than the Few and How Collective Wisdom Shapes Business, Economies, Societies and Nations*. (New York: Doubleday, 2004), introduction.
[167] Winston, *Bee Time*, 184–85.
[168] Surowiecki, *The Wisdom of Crowds,* introduction.
[169] Seeley, *Honeybee Democracy*, 8.
[170] Ibid., 118.
[171] Ibid., 173.
[172] Ibid., 101.
[173] Ibid., 142.
[174] Ibid., 211.
[175] Ibid., 99.
[176] Ibid., 145.
[177] Ibid., 218.
[178] Malcolm Gladwell, *The Tipping Point: How Little Things Can Make a Big Difference* (Boston: Little Brown, 2000), 186.
[179] Avery, *Sustainable Leadership*, 129.
[180] Gladwell, The Tipping Point, 177-180.
[181] Aver, *Sustainable Leadership*, 133.
[182] Seeley, *Honeybee Democracy*, 139.
[183] Baron, *The Desert Locust,* 121.
[184] Ibid., 39.
[185] Ibid., 39.
[186] Inayatullah Chaudhry, Sayed El Bashir, and Hassanali, "Sexual Behavior and Communication in the Desert Locust, Schistocerca Gregaria (Orthoptera: Acrididae): Sex Pheromone in Solitaria," (*Environmental Entomology*, December 1, 1994), 1544, https://doi.org/10.1093/ee/23.6.1544.
[187] Lockwood, *Locusts*, 239.
[188] Baron, *The Desert Locust*, 82.
[189] Ibid., 41.
[190] FAO Staff, FAQ.
[191] Baron, *The Desert Locust,* 95.
[192] Lockwood, *Locusts,* 148.
[193] Seeley, *Honeybee Democracy*, 219.
[194] Ibid., 74.

[195] Lockwood, *Locusts,* 256.
[196] Laura Ingalls Wilder, *On the Banks of Plum Creek* (New York: HarperCollins, 1993), 193–95.
[197] Kasey Leander, and John Stonestreet, "What Love Really Looks Like," *BreakPoint Commentary*, August 12, 2022, 1.
[198] Leander, Clive Wearing, 2.
[199] Winston, *Bee Time,* 199.
[200] Winston, *Bee Time,* 201.
[201] Seeley *Honeybee Democracy*, 9.
[202] Ibid., 10.
[203] Ibid., 12.
[204] Ibid., 11.
[205] Ibid., 123.
[206] Ibid., 176.
[207] Winston, *Bee Time*, 202.
[208] Ibid., 203-204.
[209] Staff Writer, "To honeybees, a picture is worth a thousand line angles." (*Science News*. Volume 127. No. 13. March 13, 1985), 1.
[210] Sepideh Bazazi, Jerome Buhl, Joseph Hale, Michael L. Anstey, Gregory A Sword, Stephen Simpson, and Iain D Couzin. "Collective Motion and Cannibalism in Locust Migratory Bands." (*Current Biology*, May 20, 2008), abstract..
[211] Baron, *The Desert Locust,* 120.
[212] Staff Writer, Institute of Physics, "Swarms of Locusts Use Social Networking to Communicate," (*Science Daily.Com*, July 15, 2011), www.sciencedaily.com/releases/2011/07/110714191435.htm.
[213] Institute of Physics, 2.
[214] Sir Arthur Conan Doyle, "A Scandal in Bohemia," *Adventures of Sherlock Holmes*, (Book of the Month Club: New York, 1994), 6.
[215] Frankl, *Man's Search for Meaning*, 104.
[216] Seeley, *Honeybee Democracy*, 25.
[217] Tarpy, David. "Beginner Beekeeping Training." NCSU Beekeeper Education and Engagement System, 2021.
[218] Seeley, *Honeybee Democracy*, 27.
[219] Rusty Burlew. "How to recognize robbing bees and ways to stop them." (*honeybeesuite.com*, Accessed September 19, 2023), https://www.honeybeesuite.com/robbing-bees-questions-and-answers/.
[220] Winston, *Bee Time*, 84.
[221] Ibid., 156.

[222] Chuck Eddy, "The Daddy Shady Show", *Village Voice*, Dec. 31, 2002.

[223] Lockwood, *Locusts*, 167–68.
[224] Baron, *The Desert Locust*, 33.
[225] Ibid., 33.

[226] Staff writer, "Singapore Turns Sewage into Clean, Drinkable Water, Meeting 40% of Demand", August 10, 2021. https://www.voanews.com/a/east-asia-pacific_singapore-turns-sewage-clean-drinkable-water-meeting-40-demand/6209374.html. Accessed January 23, 2024.
[227] Bazazi, Cannibalism , 735,739..
[228] Milton Leitenberg, "Deaths in Wars and Conflicts in the 20th Century," June 20, 2006, Center for International and Security Studies at Maryland, https://cissm.umd.edu/research-impact/publications/deaths-wars-and-conflicts-20th-century, accessed on January 24, 2024.
[229] Frankl, *Man's Search for Meaning*, 108.
[230] Ibid., 114.
[231] Ibid., 115.
[232] Blaise Pascal, *Pascal's Pensees*, (New York: E.P. Dutton & Co., 1958), Available on guternberg.org, 97.
[233] C C.S. Lewis, edited by Walter Hooper, *Present Concerns: Essays by C.S. Lewis*, (New York: Harcourt Brace Jovanovich, 1986), 73–80.
[234] Christopher Hitchens. "Why the suicide killers chose September 11." (*The Guardian*. October 3, 2001).
[235] History of the Battle for Vienna, https://www.youtube.com/watch?v=BKnUwWuTCXg
[236] Paul Butler, *The World and Everything In It.* Podcast News Story, September 11, 2023.
[237] Lockwood, *Locusts, 19..*
[238] Ibid., 19–20.
[239] Ibid., 20.
[240] Ibid., 163.
[241] Ibid.
[242] Ibid., 11.
[243] Ibid, 53.
[244] William J. Bennett, "The Volunteer at Auschwitz by Chuck Colson," *The Book of Virtues*. (Simon and Schuster: New York, 1993) 803-808.

Made in the USA
Monee, IL
04 November 2026

67383955R00125